Perspektiven der Mathematikdidaktik

Herausgegeben von
G. Kaiser, Hamburg, Deutschland
R. Borromeo Ferri, W. Blum, Kassel, Deutschland

In der Reihe werden Arbeiten zu aktuellen didaktischen Ansätzen zum Lehren und Lernen von Mathematik publiziert, die diese Felder empirisch untersuchen, qualitativ oder quantitativ orientiert. Die Publikationen sollen daher auch Antworten zu drängenden Fragen der Mathematikdidaktik und zu offenen Problemfeldern wie der Wirksamkeit der Lehrerausbildung oder der Implementierung von Innovationen im Mathematikunterricht anbieten. Damit leistet die Reihe einen Beitrag zur empirischen Fundierung der Mathematikdidaktik und zu sich daraus ergebenden Forschungsperspektiven.

Herausgegeben von

Prof. Dr. Gabriele Kaiser
Universität Hamburg

Prof. Dr. Rita Borromeo Ferri,
Prof. Dr. Werner Blum,
Universität Kassel

Rongjin Huang

Prospective Mathematics Teachers' Knowledge of Algebra

A Comparative Study in China and the United States of America

 Springer Spektrum

Rongjin Huang
Middle Tennessee State University
USA

Dissertation at the Texas A&M University, 2010

The book was funded by FRCAC at Middle Tennessee State University

ISBN 978-3-658-03671-3　　　　　　　　ISBN 978-3-658-03672-0 (eBook)
DOI 10.1007/978-3-658-03672-0

The Deutsche Nationalbibliothek lists this publication in the Deutsche Nationalbibliografie; detailed bibliographic data are available in the Internet at http://dnb.d-nb.de.

Library of Congress Control Number: 2014933948

Springer Spektrum
© Springer Fachmedien Wiesbaden 2014

Printed on acid-free paper

Springer Spektrum is a brand of Springer DE.
Springer DE is part of Springer Science+Business Media.
www.springer-spektrum.de

Foreword

When Rongjin Huang shared with me several months ago that his work has been accepted for publication as a monograph, I could not be happier. Rongjin conducted this study to compare knowledge of algebra for teaching (KAT) among Chinese and U.S. prospective mathematics teachers as it relates to the courses they have taken. What this book is going to share with us seems straightforward, and readers may think it is just another story about the knowledge gap between Chinese and U.S. prospective mathematics teachers. However, there are several aspects that help make this book a unique contribution, and I would like to share them with you.

First of all, this work builds upon Rongjin's years of experiences in China as a mathematics teacher, teacher educator, and researcher. Before coming to the United States in Spring 2008, Rongjin worked in China as a high school mathematics teacher and then as a teacher educator in Shanghai and Macao for several years. During his time in China, Rongjin worked very productively as a teacher and a researcher. An important feature of Rongjin's research is that he tends to develop research topics with a close connection with mathematics education practices especially at middle and high school levels. Rongjin's years of teaching experiences in high schools and universities helped him develop a long-standing interest in examining and understanding mathematics teaching, teachers' knowledge, and teacher education practices in China. Thus, readers can be assured that Rongjin brings an insider's perspective to this study about prospective mathematics teachers' knowledge and their course taking in China.

Second, Rongjin has been in the United States since the Spring of 2008, first as a doctoral student at Texas A&M University and then a mathematics teacher educator and researcher. For the past six years, Rongjin has had many opportunities to work directly with prospective teachers in various capacities. Those experiences provided him the first-hand knowledge and experience with prospective mathematics teachers and their education in the United States, which are important for designing and carrying out this cross-national comparative study.

Third, this work builds upon Rongjin's on-going research on mathematics instruction, teacher knowledge, and teacher education. His expertise is well illustrated through many of his works in these topic areas (e.g., Bednarz, Fiorentini, and Huang, 2011; Huang and Bao, 2006; Huang, Li, Zhang, and Li, 2011; Li and Huang, 2013). He has been persistent in examining and understanding the nature of mathematics instructional practices and teacher professional development, especially in China and the United States. For example, he found that Chinese teachers are more likely to implement mathematical tasks with high cognitive demands than their U.S counterparts. Moreover, Chinese teachers continuously develop their expertise in unique ways such as jointly developing public lessons. Readers can thus expect to gain further insights from this book about cross-national similarities and differences in mathematics knowledge for teaching among Chinese and U.S. prospective mathematics teachers.

It needs to be clarified that Rongjin's background and on-going research are not enough to justify, but help position him to make the unique contribution to this book. Here I want to highlight another three aspects that help distinguish this book from what can be easily perceived as another 'trivial' comparative study.

1. This book presents cross-national similarities and differences in KAT among Chinese and U.S. prospective mathematics teachers, which also helps contribute to our understanding of the nature of KAT from an international perspective.

This book focuses on a study that examines similarities and differences in knowledge of algebra for teaching (KAT) between Chinese and U.S. prospective mathematics teachers. Readers can certainly learn about the reported knowledge gap between these two groups of prospective mathematics teachers; a result that has been speculated but not studied before.

In my view, the unique value of this work goes beyond reporting the knowledge gap to make potential theoretical contributions to our understanding of the nature of KAT from an international perspective. It is commonly acknowledged that teachers' knowledge is key to effective classroom instruction, but less agreed on specific knowledge components and their structure. Rongjin presented and discussed several different perspectives about mathematics teachers' knowledge in general, algebra knowledge for teaching in specific. These perspectives are valuable in certain ways, but are typically derived from works within a specific education system and cultural context. Understanding the nature of mathematics teachers' knowledge can be strengthened from an international perspective (Li, 2014). Rongjin made a unique contribution through this study by examining and discussing different knowledge components (i.e., school mathematics, advanced mathematics, and teaching mathematics) and the structure of KAT based on the data obtained from Chinese and U.S. prospective mathematics teachers. The relationships of these knowledge components are clearly not universal but vary by several factors embedded in different program and cultural contexts.

2. This book reports cross-national similarities and differences in KAT with the use of a modified survey, which also helps contribute to our understanding of its feasibility and limitations.

The results reported in this book were obtained by surveying selected prospective mathematics teachers in China and the United States. Questions can easily be raised about the study's sampling method (376 prospective mathematics teachers from five universities in China but 115 from only one university in the United States) and the development and use of a modified survey instrument. It is clear that Rongjin kept these limitations in mind, and used alternative approaches that help contribute to our understanding of its feasibility and limitations. In particular, to address the sampling limitation, he classified the five participating universities in China into three categories in terms of their institutions' ranking. This classification helps provide differentiations and clarity about prospective mathematics teachers' KAT in his data analyses and reporting.

Moreover, the survey instrument was adapted from one that was originally developed and used in a U.S. project. The use of such a U.S.-originated survey instrument in an international comparative study requires further validation and possible modifications. As explained in this book, Rongjin took necessary steps to modify the survey instrument based on a study of mathematics curriculum standards in China, experts' validation of the survey's content, and Chinese in-service teachers' validation of the survey instrument. Moreover, he also added some open-ended items that helped elicit prospective teachers' thinking behind their answers, as well as follow-up interviews with selected prospective U.S. teachers who provided very limited explanations. All of these steps illustrate some unique features in the research methodology that Rongjin employed and contributed in this seemingly straightforward international study.

3. This book relates to cross-national similarities and differences in mathematics training provided to prospective mathematics teachers, which also helps contribute to our understanding of cultural values and what is important for prospective mathematics teachers to learn.

The results present a picture that is consistent with previous studies about the knowledge gap between Chinese and U.S. in-service teachers (e.g., Ma, 1999).

Readers should not be surprised to find out that prospective mathematics teachers' KAT relates to what they are expected to learn in and through teacher preparation programs (i.e., opportunity-to-learn, or OTL). As presented in this book, the number of courses in mathematics content and mathematics education has dramatic differences across teacher preparation programs in China and the U.S. While teacher preparation programs in China tend to require many more mathematics content courses than those programs in the U.S., the pedagogical training required for prospective teachers presents a reversed picture. Rongjin pointed out that similar differences were also reported in previous studies. Such consistent differences further suggest a cultural root in different valuations of what are important for prospective mathematics teachers to learn between China and the United States (Lappan and Li, 2002).

Like many other studies, this book not only provides readers with many important and unique results, but it also leaves readers with many more questions unanswered. With some questions highlighted at the end of this book, I am sure that reading this book will inspire readers to come up with more questions either directly related to this study or to teachers' knowledge and their education in one's own education system. With this belief, I recommend this book as an important reading to all of those who are serious about examining teachers' mathematics knowledge for teaching and exploring ways of improving teachers' professional development.

Yeping Li
Texas A&M University
College Station, USA
yepingli@tamu.edu

References:

Bednarz, N., Fiorentini, D., & Huang, R. (Eds.). (2011). International approaches to professional development for mathematics teachers. Canada: Ottawa University Press.

Huang, R., & Bao, J. (2006). Towards a model for teacher's professional development in China: Introducing keli. Journal of Mathematics Teacher Education, 9, 279-298.

Huang, R., Li, Y., Zhang, J., & Li, X. (2011). Developing teachers' expertise in teaching through exemplary lesson development and collaboration. ZDM- The International Journal on Mathematics Education, 43(6-7), 805-817.

Lappan, G., & Li, Y. (2002). Reflections and recommendations. International Journal of Educational Research, 37, 227-232.

Li, Y. (2014). Learning about and improving teacher preparation for teaching mathematics from an international perspective. In S. Blömeke, F.-J. Hsieh, G. Kaiser & W. H. Schmidt (Eds.). International perspectives on teacher knowledge, beliefs and opportunities to learn. (pp. 49-57). Dordrecht: Springer.

Li, Y., & Huang, R. (Eds.) (2013). How Chinese teach mathematics and improve teaching. New York: Routledge.

Acknowledgments

My heartfelt thanks goes to my Chair, Dr. Yeping Li, for his insightful academic guidance and the sharing of his experience and expertise. I would like to thank my Co-chair, Dr. Gerald Kulm, for his strong support both academically, financially, and his many hours of editing and commenting to improve this dissertation. I wish to thank my committee members for their support during the study. In particular, Dr. Willson has given me a great amount of help in developing my capability of using quantitative methods. Dr. Donal Allen has provided insights in developing the questionnaire from a mathematician perspective.

I also thank Dr. Dennie Smith for the strong support of conference travels and data collection. I appreciate Dr. Dianne Goldsby and Dr. Dawn Parker for their help in data collection and allowing me to observe their excellent teaching. I would like to thank Ms. Sandra Nite for her help in gaining experience in teacher professional development in the U.S.

Thanks also goes to my friends, Dr. Jiansheng Bao, Dr. Xuemei Chen, Dr. Meiyue Jin, Dr. Zhongru Li, Dr. Hongyu Su, Professor Liyun Ye, Professor Linquan Wang, and Professor Weizong Zhu, for their valuable help in collecting data in China. I thank Dr. R. E. Floden at Michigan State University for allowing the use of their instrument of KAT (Knowledge of algebra for teaching) and sharing his expertise. I also thank Dr. Jinfa Cai at University of Delaware, Dr. Shuhua An and Dr. Xuehui Li at California State University – Long Beach, for sharing their expertise and their willingness with the assistance of the data collection. I would like to express my appreciation to the prospective teachers who participated in this study for their time and efforts to complete the surveys.

I also want to extend my gratitude to Mr. Kyle Prince and Mrs. Teresa Schmidt from Middle Tennessee State University for their help in editing the manuscript, and a Faculty Research and Creative Activity Grant from Middle Tennessee State University for funding the publication of this dissertation.

Finally, I am grateful to my wife, Lifeng Jin, for her sacrifice, support, and love. I also thank my lovely daughter, Shuying Huang, for her love, understanding, and help.

December 20, 2013, Murfreesboro, TN

Rongjin Huang

Table of Contents

Figures

Tables

Nomenclature

MKT   Mathematics Knowledge for Teaching

KAT   Knowledge of Algebra for Teaching

KTCF   Knowledge for Teaching the Concept of Function

1 Chapter One: Introduction

1.1 Background

Preparing future mathematics teachers with the appropriate mathematics knowledge needed for teaching is crucial for high quality teaching, eventually resulting in student learning (National Mathematics Advisory Panel [NMAP], 2008; RAND Mathematics Study Panel, 2003; Rowland & Ruthven, 2010; Sullivan & Wood, 2008). Researchers have focused on identifying and measuring *mathematics knowledge for teaching* (denoted as MKT) (e.g., Ball, Hill, & Bass, 2005; Blömeke & Delaney, 2012; Depaepe, Verschaffel, & Kelchtermans, 2013; Hill, Ball, & Schilling, 2008; Kulm, 2008; McCrocy, Floden, Ferrini-Mundy, Reckase, & Senk, 2012; Rowland & Ruthven, 2010; Tatto et al., 2012). For example, drawing on Shulman's (1986) seminal work on teacher knowledge, Ball and her colleagues have developed a refined framework of MKT and relevant instruments for measuring MKT (Ball et al., 2005; Ball, Thames, & Phelps, 2008), which previously focused on elementary school teachers, and recently extended to middle school teachers. Moreover, researchers found that MKT has a close relationship with classroom instruction (Hill, Blunk et al., 2008) and student achievement (Hill, Rowan, & Ball, 2005; Baumert et al., 2010).

As mathematics literacy - particularly algebra - became an extension of the civil rights movement (Dubinsky & Moses, 2011; Moses & Cobb, 2001), teaching algebra to all students (Edward, 1990) has revitalized its importance (Creenes & Rubenstein, 2008; National Council of Teachers of Mathematics [NCTM], 2009; NMAP, 2008). While studies have focused on on the teaching and learning of algebra for decades (e.g., Kaput, Blanton, & Moreno, 2008; Katz, 2007; Kieran, 1992, 2007; Dubinsky & Wilson, 2013), less attention has been given to teachers' knowledge of algebra for teaching (e.g., Doerr, 2004; Even, 1993; Even

& Tirosh, 1995, 2008; McCrory et al., 2012). In order to understand and develop teachers' *Knowledge of Algebra for Teaching* (denoted as KAT), a research team at Michigan State University (Ferrini-Mundy et al., 2006; McCrory et al., 2012) developed a framework for identifying KAT and an instrument for measuring KAT. The framework of KAT (McCrory et al., 2012) included *School Algebra Knowledge* (i.e., algebra in secondary school, denoted as SA), *Advanced Mathematical Knowledge* (i.e., related college math such as calculus and abstract algebra, denoted as AM), and *Mathematics–for-Teaching knowledge* (i.e., knowledge of typical errors, canonical uses of school math, curriculum trajectories, etc., denoted as TM). An instrument grounded in this model has been developed and tested in the U.S. with an internal consistency Cronbach's alpha .8 (Floden & McCrory, 2007; Floden, McCrory, Reckase, & Senk, 2009).

Efforts to pursue high quality classroom teaching and student learning in mathematics have led researchers to explore the practices in high-achieving countries, such as China (e.g. Mullis et al., 2008; OECD, 2010). A large number of comparative studies of mathematics education between China and the United States have covered a broad range of topics in mathematics education. These studies include student learning (e.g., Cai, 1995, 2000, 2004), classroom teaching (e.g., Huang & Cai, 2010; Huang & Li, 2010; Stevenson, Chen, & Lee, 1993; Stevenson & Lee, 1995), teachers' knowledge (e.g., An, Kulm, & Wu, 2004; Ma, 1999) and beliefs (i.e., An, Kulm, Ma, & Wang , 2006; Cai, 2000, 2006; Cai, Perry, Wong, &Wang, 2009), and curriculum (i.e., Fan & Zhu, 2007; Kulm & Li, 2009; Li, Chen, & An, 2009).

With regard to teachers' knowledge, Ma (1999) found that Chinese elementary mathematics teachers demonstrated a profound understanding of fundamental knowledge for teaching in contrast to their U.S. counterparts. A recent study on mathematics teacher preparation at the middle school level in Chinese Taiwan, South Korea, Bulgaria, Germany, Mexico and the United States, found that "in Chinese Taiwan and Korea, the level of mathematics preparation was very strong and in both countries, the amount of emphasis given to the practical issues of mathematics pedagogy was also extensive" (Schmidt et al., 2007, p. 1). In the Teacher Education Development Study in Mathematics (TEDS-M) (Tatto et al., 2012), knowledge for teaching mathematics included mathematics content knowledge (MCK) and mathematics pedagogical content knowledge (MPCK). Among the 17 participant countries, Chinese Taiwan again performed outstanding in both MCK and MPCK. In a study comparing future teachers' professional knowledge of elementary mathematics from an advanced standpoint in China, Korea and Germany, Buchholtz, Leung, Ding, Kaiser, Park and Schwarz (2013) found that both China and Korea participants outperformed Germany counterparts.

By analyzing the curriculum designs of mathematics teacher preparation programs in Chinese Mainland and Korea, Li, Huang, and Shin (2008) revealed that the secondary teacher (including middle and high school levels) preparation programs in the Chinese Mainland and Korea emphasize teachers' learning of mathematics subject matter knowledge.

In addition, a study comparing pedagogical content knowledge of middle school mathematics teachers between the U.S. and China (An et al., 2004) has found that the Chinese mathematics teachers emphasized gaining correct conceptual knowledge by relying on more rigid development of procedures. The U.S. teachers emphasized a variety of activities designed to promote creativity and inquiry to develop concept mastery, with a lack of connection between manipulative and abstract thinking, and between understanding and procedural development.

Previous studies described some features of mathematics teacher knowledge for teaching in China and the U.S. at elementary and middle school levels, but there is no comparative study on teachers' knowledge at high school level. Specifically, there is no comparative study focusing on teachers' knowledge for teaching specific content at secondary school level. This study is designed to compare teacher knowledge for teaching algebra in China and the U.S. to address the deficiency.

Teacher preparation in China attempts to prepare secondary mathematics teachers (including middle and high school levels) with a solid mathematics foundation and broad mathematics background, with less attention paid towards pedagogical knowledge preparation (Li et al., 2008). Compared to middle school mathematics teacher preparation in East Asia, the practice in the U.S. seems to place less emphasis on mathematics content knowledge and pedagogical content knowledge, but spend more time on learning pedagogical knowledge in general

(Babcock et al., 2010; Schmidt et al., 2007; Tatto et al., 2012). The differences between the teacher preparation systems in the U.S. and China may result in differences of teachers' knowledge for teaching. In this study, I aim to examine the status and characteristics of mathematics teachers' knowledge by focusing on their knowledge of algebra for teaching, and further investigate the relationship between teachers' knowledge for teaching and their courses taken.

Adapting an instrument cross-culturally is a challenging and important issue in comparative studies (Delaney, Ball, Hill, Schilling, & Zopf, 2008). In this study, I developed a questionnaire measuring mathematics knowledge of algebra for teaching (KAT) based on an existing instrument developed by Michigan State University (Floden et al., 2009), and used it to collect prospective teachers' data in China and the U.S. Comparing the features of KAT between China and the U.S. could broaden and deepen our understanding of mathematics knowledge of algebra for teaching. In addition, this study could contribute to developing and validating a survey instrument of KAT cross-culturally. Thus, this study will provide insights into the understanding of mathematics knowledge of algebra for teaching, the development of instruments of KAT, and the improvement of secondary mathematics teacher preparation.

Although teaching algebra for all has been a slogan of mathematics education reformers (Edwards, 1990) for two decades, it is still a very challenging task for the current mathematics education reform (Katz, 2007; Kieran, 2004, 2007; NCTM, 2000, 2009; CCSSI, 2010). As described previously, there are salient differences between the U.S. and Chinese Mainland or Chinese Taiwan mathematics teacher preparation programs in middle and secondary schools in terms of their emphasis of subject matter knowledge and pedagogical content knowledge (Li et al., 2008; Schmidt et al., 2007; Tatto et al., 2012). However, there is no empirical study to examine similarities and differences in teachers' knowledge for teaching at secondary school level in China and the U.S.

1.2   Statement of Purpose

The purpose of this study is to examine prospective teachers' *Knowledge for Algebra Teaching* (KAT) in China and the U.S. by using a mixed research method. In particular, I will compare teachers' KAT between these two countries at an item level and a structure level. At the item level, I mainly focused on mean differences. At the structure level, a SEM model (Structural Equation Model) was used to conduct a path model and measurement model analysis cross-culturally. In addition, teachers' *Knowledge for Teaching the Concept of Function* (KTCF) was investigated qualitatively.

1.3   Research Questions

The main purpose is to explore the characteristics of prospective teacher knowledge of algebra for teaching in China and the U.S. However, I also realize that the features of programs that participants attended should have an impact on their performance of KAT. Thus, the number of courses taken is selected as a key factor, which may have direct effect on teachers' KAT. In particular, the study is aimed to answer the following research questions:

1. What are the differences and similarities of KAT between Chinese and U.S. prospective teachers?

2. What are the relationships among different components of KAT within and between China and the U.S.?

3. What are the differences and similarities between Chinese and U.S. prospective teachers' *Knowledge for Teaching Concept of Function* (KTCF)?

4. What are the relationships between prospective teachers' status of KAT and their course- taking?

1.4   Delimitations

This study only examines prospective teachers' knowledge of algebra for teaching. Since algebra topics are mainly included in secondary (i.e., middle and high school) mathematics, I only focus on the population of prospective secondary school teachers, not on elementary teachers.

2 Chapter Two: Literature Review

Recently, an effort has been made to measure mathematics teachers' knowledge needed for teaching even though there is more than 20 years of history in studying teachers' knowledge. Shulman's (1986) classification of subject matter knowledge, pedagogical knowledge, and curriculum knowledge laid a foundation for the study of teacher knowledge. Drawing on Shulman's framework, researchers have further refined and developed models to better describe and measure teacher knowledge needed for teaching (e.g., Ball et al., 2005; Krauss et al., 2008; Schmidt et al., 2007; Silverman & Thompson, 2008; Tatto et al., 2012). This literature review consists of two sections: teacher knowledge needed for teaching, and mathematics preparation of teachers in China and the U.S. In the first section, I reported the conceptualization of teacher knowledge for teaching in general and discussed the models for describing teacher knowledge for teaching mathematics in particular. Then, I discussed relevant studies on teachers' knowledge needed for teaching algebra. In the second section, I analyzed the mathematics education systems and mathematics teacher preparation in China and the U.S., and summarized relevant studies on mathematics teachers' knowledge for teaching in China and the U.S. Finally, grounded in the literature review, a framework for this study was proposed.

2.1 Knowledge Needed for Teaching

Great efforts have been made to seek what kind of knowledge a teacher needs to know in order to teach students effectively. In Shulman's (1986) seminal work on teachers' knowledge, he identified three categories, namely, *content knowledge, curriculum knowledge,* and *pedagogical knowledge.* The first, *content knowledge,* includes knowledge of the subject and its organizing structures. The teacher needs not only to understand *that* something is so; the teacher must further understand *why* it is so. The second, *curricular knowledge,* is "represented by the full range of programs designed for the teaching of particular subjects and topics at a given level, the variety of instructional materials available in relation to those programs, and the set of characteristics that serve as both the indications and contraindications for the use of particular curriculum or program materials in particular circumstances" (p. 9). The third, *pedagogical content knowledge* (PCK), is described as follows:

> The most useful forms of representation of those ideas, the most powerful analogies, illustrations, examples, explanations, and demonstrations – in a word, the most useful ways of representing and formulating the subject that makes it comprehensible to others. Pedagogical content knowledge also includes an understanding of what makes the learning of specific topics easy or difficult: the conceptions and preconceptions that students of different ages and backgrounds bring with them to the learning of those most frequently taught topics and lessons. (p. 9)

Since Shulman (1986) coined the term PCK, many researchers have attempted to illustrate and clarify the nature of PCK and its implications for teacher education (e.g., Gess-Newsome, 1999). However, pedagogical content knowledge is often not clearly distinguished from other forms of teacher knowledge. For example, pedagogical content knowledge has been defined as "the intersection of knowledge of the subject with knowledge of teaching and learning" (Niess, 2005, p. 510) or as "that domain of teachers' knowledge that combines subject matter knowledge and knowledge of pedagogy" (Lowery, 2002, p. 69). An even more careful description of PCK is still unclear:

> Pedagogical content knowledge is a teacher's understandings of how to help students understand specific subject matter. It includes knowledge of how particular subject matter topics, problems, and issues can be organized, represented and adapted to the diverse interests and abilities of learners, and then presented for instruction. The defining feature of pedagogical content knowledge is its conceptualization as the result of a *transformation* of knowledge from other domains (Magnusson, Krajcik, & Borko, 1999, p. 96).

In summary, PCK includes multiple components: (1) knowledge of students' (mis)conceptions and difficulties, (2) knowledge of instructional strategies, (3) knowledge of mathematical tasks and cognitive demands, (4) knowledge of educational ends, (5) knowledge of curriculum and media, (6) context knowledge, (7) content knowledge, and (8) pedagogical knowledge (Depaepe, Verschaffel, & Kelchtermans, 2013).

2.2 Mathematics Knowledge for Teaching

There is a widespread agreement that mathematics teachers need to have a deep understanding of mathematics (Ball, 1993; Grossman, Wilson, & Shulman, 1989; Ma, 1999). However, teachers' knowledge of mathematics alone is insufficient to support their attempts to teach mathematics effectively. There are various ways to define PCK in mathematics. While Ball (1990) differentiated two dimensions of teachers' content knowledge: teachers' ability to execute an operation (division by a fraction) and their ability to represent that operation accurately for students, Ma (1999) described "profound understanding of fundamental mathematics" in terms of the connectedness, multiple perspectives, fundamental ideas, and longitudinal coherence. Moreover, the National Research Council [NRC] suggested that mathematics teachers need specialized knowledge that "includes an integrated knowledge of mathematics, knowledge of the development of students' mathematical understanding, and a repertoire of pedagogical practices that take into account the mathematics being taught and the students learning it." (Kilpatrick, Swafford, & Findell, 2001, p. 428)

A research team at the University of Michigan has focused on understanding and measuring mathematical knowledge that was considered pertinent and important for teaching (Ball & Bass, 2000; Ball et al., 2005; Hill, Ball, & Schilling, 2008). Furthermore, Hill and her colleagues further explored the relationship between mathematics knowledge needed for teaching (MKT) and students' achievement (Hill et al., 2004), and classroom instruction (Hill, Ball, & Schilling, 2008). Growing attention has been given to studying characteristics of teacher's knowledge needed for teaching specific content areas (e.g., Ball et al., 2005; Even, 1990, 1993; Ferrini-Mundy et al., 2006; Ma, 1999), which will be discussed in the following sections.

Researchers have attempted to understand *what* mathematical knowledge is entailed in teaching, how to assess it (Ball & Bass, 2000; Ball et al., 2005; Hill, Schilling, & Ball, 2004), and how to develop and refine ways to effectively promote *mathematical knowledge for teaching* (MKT) in teacher education and teacher professional development programs (NMAP, 2008; Stylianides & Stylianides, 2006). Ball and her colleagues have developed a specific framework describing MKT (Ball et al., 2005). According to this model, subject matter knowledge is divided into two categories: *Common Content Knowledge* (CCK), which can be developed in anyone who has had school mathematics education, and *Specialized Content Knowledge* (SCK), which is used mainly by teachers. Meanwhile, the model makes a distinction between two main categories within pedagogical content knowledge: *Knowledge of Content and Students* (KCS) and *Knowledge of Content and Teaching* (KCT). This model highlights the kind of mathematical content knowledge that is the specialty of teachers, and recognizes that knowledge of mathematics for teaching is partially the product of content knowledge interacting with students in their learning processes and with teachers in their teaching practices.

Grounded in the concept of mathematics proficiency (Kilpatrick et al., 2001), Kilpatrick, Blume and Allen (2006) proposed a framework for Mathematical Proficiency for Teaching. It suggests that *mathematical proficiency with content* (MPC) and *mathematical proficiency in teaching* (MPT) should be the main components for teachers to teach for mathematics proficiency. The *mathematical proficiency with content* (MPC) includes conceptual understanding, procedural fluency, strategic competence, adaptive reasoning, productive disposition, cultural and historical knowledge, knowledge of structure and conventions, and knowledge of connections within and outside the subject. The *mathematical proficiency in teaching* (MPT) consists of knowing students as learners, assessing one's teaching, selecting or constructing examples and tasks, understanding and translating across representations, understanding and using classroom discourse, knowing and using the curriculum, and knowing and using instructional tools and materials. This model illustrates Shulman's (1986) subject matter knowledge and pedagogical knowledge with a focus on mathematics proficiency.

Simon (2006) adopted the idea of a *Key Developmental Understanding* (KDU) in mathematics, namely, understanding a topic from multiple perspectives, building a well-structured knowledge web surrounding the topic as a way to think about understandings. KDUs are regarded as powerful springboards for learning and useful goals of mathematics instruction. Silverman and Thompson (2008) argued that developing MKT involves transforming these personal KDUs of a particular mathematical concept to an understanding of: (1) how this KDU could empower their students' learning of related ideas; (2) actions a teacher might take to support students' development of KDU and reasons why those actions might work. They further suggested a framework of mathematical knowledge for teaching as follows: A teacher has developed knowledge that supports conceptual understanding of a particular mathematical topic when he or she (1) has developed a KDU within which that topic exists, (2) has constructed models of the variety of ways students may understand the content, (3) has an image of how someone else might come to think of the mathematical idea in a similar way, (4) has an image of the kinds of activities and conversations about those activities that might support another person's development of a similar understanding of the mathematical idea, and (5) has an image of how students who have come to think about the mathematical idea in the specified way are empowered to learn related mathematical ideas. This framework opens up the possibility for the goal of mathematics teacher education to shift from positioning prospective teachers to develop particular MKT to developing professional practices that would support teachers' abilities to continually develop MKT.

The Teacher Education Development Study in Mathematics (TEDS-M) (Tatto et al., 2012) identified two components of MKT: *Mathematics content knowledge* (MCK) and *mathematics pedagogical content knowledge* (MPCK). MCK, in the TEDS-M framework, includes not only basic factual knowledge of mathematics but also the conceptual knowledge of structuring and organizing principles of mathematics as a discipline. MPCK consists of three domains: mathematics curricular knowledge, knowledge of planning, and knowledge for enacting mathematics.

The COACTIV (Professional Competence of Teachers, Cognitively Activating Instruction, and the Development of Students' Mathematical Literacy) project in Germany (Baumert et al., 2010; Krauss et al., 2008) examined the relationship between content knowledge (CK) and pedagogical content knowledge (PCK). It was found that although CK and PCK are highly correlated, they are statistically distinct and CK alone is not sufficient, but a necessary condition for PCK. In addition, it was noticed that CK and PCK are more correlated in teachers who had a more thorough education in mathematics (Krauss et al., 2008) and instructional quality is significantly more correlated to PCK than CK (Baumert et al., 2010).

The previously described studies enrich and/or extend Shulman's taxonomy with a focus on mathematics subject, and provide insights into the construct of CK and PCK and the relationship between them. In the next section, specific research on teacher's knowledge needed for teaching algebra will be analyzed.

2.3   Teachers' Knowledge of Algebra for Teaching

Algebra is an important part of school mathematics but is challenging for students to learn (Blume & Heckman, 1997; NCTM, 2000; NMAP, 2008). As more students take algebra, more teachers teach algebra, yet the preparation of algebra teachers has not been widely researched (Stein, Kaufman, Sherman, & Hillen, 2011). Very little research has examined teachers' knowledge of algebra for teaching (Artigue, Assude, Grugeon, & Lenfant, 2001; Even, 1990, 1993; Ferrini-Mundy et al., 2006; Li, 2007). Artigue and colleagues differentiated three dimensions of knowledge of algebra for teaching as follows: (1) epistemological dimension; (2) cognitive dimension; and (3) didactic dimension. The *epistemological dimension* includes: (a) the complexity of the algebraic symbolic system and the difficulties of its historical development, and (b) how to flexibly use algebraic tools in solving different kinds of problems that are internal or external to the field of mathematics. The *cognitive dimension* deals with knowledge about learning processes in algebra, which includes knowing (a) the development of the student's algebraic thinking and (b) students' interpretations of algebraic concepts and notations. The *didactic dimension* involves knowledge of (a) the algebra curriculum, and (b) the specific goals of algebraic teaching at a given grade.

Even (1990) identified and illustrated seven dimensions of subject matter knowledge based on an in-depth examination of the concept of function: (1) essential features, (2) different representations, (3) alternative ways of approaching, (4) the strength of the concept, (5) basic repertoire, (6) knowledge and understanding of a concept, and (7) knowledge about mathematics. *Essential features* are referred to as concept images by Vinner (1983) as the mental pictures of this concept, together with the set of properties associated with the concept (in the person's mind). It is crucial for teachers to judge if an instance belongs to a concept family by using an analytical judgment as opposed to a mere use of a prototypical judgment. It is necessary that teachers are able to correctly distinguish between concept examples and non-examples. *Different representations* give different insights which allow a better, deeper, more powerful and more complete understanding of a concept. When dealing with a mathematical concept in different representations, one may abstract the concept by grasping the common properties of the concept while ignoring the irrelevant characteristics that are imposed by the specific representation at hand. *Alternative ways of approaching the same concept* are used to deal with complex concepts in various forms, representations, labels and notations. *The strength of the concept* refers to the importance or power to open new possibilities, understand new concepts and capture the essence of the definition, as well as a more sophisticated formal mathematical knowledge. *Basic repertoire* includes powerful examples that illustrate important principles, properties, and theorems. The basic repertoire should be well known and familiar in order to be readily available for use. *Knowledge and understanding of a concept* means to achieve procedural proficiency and conceptual knowledge. The learning of a new concept or relationship implies the addition of a node or link to the existing cognition structure; thus making the whole more stable than before. *Knowledge about mathematics* includes knowledge about the nature of mathematics. This is a more general knowledge about a discipline, which guides the construction and use of conceptual and procedural knowledge.

Comparing Artigue et al. (2001) and Evens' (1990) models, categories (1), (4), (6) and (7) of Even's category belong to the epistemological dimension while the others belong to the didactic dimension.

Ferrini-Mundy and her colleagues (2006, 2012) have developed a two-dimensional framework that describes mathematics knowledge of algebra for teaching. In their model, the horizontal dimension indicates the fundamental *categories of knowledge* involved in teaching algebra, while the vertical dimension identifies several *tasks of teaching* in which teachers may apply their mathematical knowledge. The three overarching categories, *decompressing*, *trimming*, and *bridging*, are more sophisticated mathematical practices that utilize multiple elements of knowledge of algebra for teaching and involve multiple tasks of teaching. *Categories of knowledge* include core content knowledge, representation, content trajectory, application and context, language and convention, and mathematical reasoning and proof. *Tasks of teaching* consist of analyzing students' work and thinking, designing, modifying and selecting mathematical tasks; establishing and revising mathematical goals for students; accessing and using tools and resources for teaching; explaining mathematical ideas and solving mathematical problems; building and supporting mathematical community and discourse. The framework illustrates the overall landscape of knowledge of algebra for teaching: the major types of knowledge that may be used and contexts in which they may be used. *Decompressing* refers to the decompression of knowledge in the practice of teaching. For secondary school algebra teachers, what needs to be decompressed may include algorithms for solving equation, systems of equations, for simplifying expressions, and for moving among representations. *Trimming* refers to forming the mathematical content in a way that matches students' current levels of sophistication while treating the mathematics with integrity (McCrory et al., 2012). *Bridging* includes connecting and linking mathematics across topics, courses, concepts, and goals, including connecting the ideas of school algebra to those of abstract algebra and real analysis, linking one area of school mathematics to another (McCrory et al., 2012).

Flodden and McCrery (2007) have further created a three dimensional construct, as illustrated in Figure 2.1, to guide the measuring of teachers' knowledge of algebra for teaching.

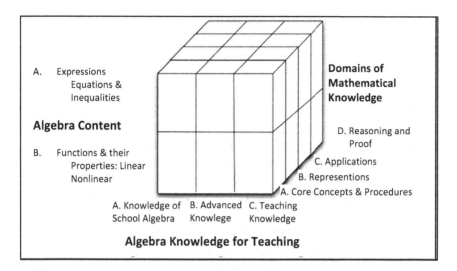

Figure 2.1. A framework for assessing knowledge of algebra for teaching.

In this framework, the base of the matrix consists of three types of algebra knowledge for teaching including knowledge of school algebra, advanced algebra knowledge and teaching knowledge. *School Algebra Knowledge* refers to the algebra covered in the curriculum from K-12. *Advanced Algebra Knowledge* includes other mathematical knowledge such college-level mathematics that provides perspectives on the trajectory and growth of mathematical ideas beyond school algebra. Typically, those areas of mathematics are found in calculus, linear algebra, real and complex analysis, and mathematical modeling. Similar to specialized content knowledge in the Hill et al. (2008) model, *Teaching Knowledge* includes "knowledge that is mathematical, that is intuitively useful for teaching, and that is unlikely to be taught explicitly, except to teachers" (McCrory et al., 2012, p. 608). Teaching knowledge also includes typical errors, canonical uses of school math, and curriculum trajectories among others. There are four *domains of mathematical knowledge* or aspects of algebra teaching and learning (core concepts and procedures, representations, applications and contexts, and reasoning and proof on the Y-axis). The Z-axis contains two major themes in school *algebra content*: expressions, equations and inequalities, and functions and their properties. Assessment items can be specifically written for each cell in the matrix; for instance, knowledge of school algebra that is related to a core procedure for solving equations. Each assessment item would be uniquely located in Figure 2.1 as a coordinated system.

Based on the existing frameworks, Li (2007) reported a refined framework for investigating teachers' mathematical knowledge of algebra for teaching linear equations. His framework consists of three domains of knowledge: knowledge of the subject matter, knowledge of learners' conceptions, and knowledge of didactic representations. *Knowledge of the subject matter* refers to mathematics subject matter as a system of established definitions, properties, facts, relations and connections; use of notations and representations; and methods for reasoning and problem solving. *Knowledge of learners' conceptions* includes the subject matter as understood by learners, including typical pre-conceptions, misconceptions, mistakes, questions, difficulties, strategies, reasoning, and factors that make a particular concept or procedure easy or difficult. *Knowledge of didactic representations* refers to how the subject matter is unpacked, linked, organized, and tailored through purposeful sequencing of topics and choices of examples, models, explanations, tasks, metaphors, and technological presentations.

2.4   Mathematics Knowledge for Teaching Some Key Concepts in Algebra

Recent research has examined teachers' knowledge needed for teaching several important concepts in school algebra, such as function, expressions, and equations.

2.4.1 Teaching and Learning of the Concept of Function

There have been different approaches to developing meaningful algebra (i.e., Bednarz, Kieran, & Lee, 1996; Hart, 1981; Usiskin, 1988). Usiskin (1988) summarized the following four approaches: (1) algebra as generalized arithmetic, (2) algebra as a study of procedures for solving certain kinds of problems, (3) algebra as the study of relationships among quantities, and (4) algebra as the study of structures. Learning algebra should include the following three core activities: generational, transformational and global/meta-level (Kieran, 2004). Function concept is one of the most important but difficult concepts across middle and high school levels (NCTM, 2000, 2006). Based on the theory of process-object duality of mathematics concept development (Sfard, 1991, 1992), a model of developing function concept is described as four stages: pre-function, action, process and object (Briedenbach, Dubinsky, Hawks, & Nichols, 1992). According to this model, for *pre-function* indicates that the student does not display very much of a function concept. An *action* is the repeatable mental or physical manipulation of an object; such a conception of function would involve, for example, the ability to substitute numbers into an algebraic expression and evaluate. *Process* involves a dynamic transformation of quantities according to some repeatable means that, given the same original quantity, will always produce the same transformed quantity. A function is conceived as an *object*, if it is possible to perform action on it, namely transformations.

In general, three representations are used for presenting functions: (1) geometrical representations including chart, graph, and histogram; (2) numerical representations including numbers, table, and ordered number pairs; and (3) algebraic representations including letters, formula, and mapping (Verstappen, 1982). However, different representations play different roles in helping students understand the concept of function (Schwartz & Yerushalmy, 1992). For example, the algebraic representations benefit the understanding of function as a *process*, while the graphical representations help in understanding functions as an *object*. Moreover, certain operations are easier to understand in particular representations. For example, composition functions performed on algebraic representations are easy to understand, while transformations performed on graphical representations will be much easier to understand. Thus, it is critical to select appropriate representations with regard to different contexts.

There are a large number of studies on teachers' knowledge for teaching the concept of function (e.g., Even, 1990, 1992, 1993, 1998; Norman, 1992, 1993). For example, based on a study on 10 secondary teachers' knowledge of the function concept, Norman (1992) found that the secondary teachers tended to have *inflexible images* of the concept of function that restricted their abilities to identify functions in unusual contexts and to shift among representations of functions. The sampled teachers were able to give formal definitions of function, distinguish functions from relations, and correctly identify whether or not a given situation was functional. However, the teachers did not show strong connections between their informal notions of function and formal definitions, and were not comfortable with generating contexts for functions. Hitt (1994) investigated 117 mathematics teachers' ideas on function and found that the teachers had difficulty in constructing functions that were not continuous or piece-wise defined. Consistent with Norman's findings, Chinnappan and Thomas (2001) found that all four prospective secondary teachers in their study had a preference of thinking about function graphically, had a weak understanding of the connections of representations, and a limited ability to describe applications of functions.

Based on a survey of teachers' knowledge about function with 152 prospective secondary teachers, Even (1993) found that many prospective secondary teachers did not hold a modern conception of a function as a univalent correspondence between two sets. These teachers tended to believe that functions are always represented by equations and that their graphs are well behaved. None of the teachers had a reasonable explanation of the need for functions to be univalent and overemphasized the procedure of the "vertical line test" without concern for understanding. Given the often weak and fragile understanding of secondary mathematics teachers about the concept of function, it is not surprising to find that the knowledge of an experienced fifth grade teacher was missing several key ideas (such as univalence and unclear notions of dependency) and lacked a notion of the connectivity among representations (Leinhardt, Zaslavsky, & Stein, 1990; Stein, Baxter, & Leinhardt, 1990).

In summary, teachers have difficulties (1) in understanding the concept of functions as a univalent correspondence between two sets (sometimes, it is not presented by a formula, or it is not of a discontinuous graph), (2) shifting different representations flexibly, and (3) relating formal function notion to contextual situations that produce the function.

2.4.2 Teaching and Learning of Expressions and Equations

Expressions. An algebraic expression can be seen as a string of symbols, a computational process, or a representation of a number. An expression can also become a function representing change if the context changes (Sfard & Linchevski, 1994). Sfard and Linchevski also clarify a potential issue in understanding algebraic expressions when they note:

> ...the difficulty lies ... in the necessity to imbue the symbolic formulae with the double meanings: that of computational procedures and that of the objects produced.... To those who are well versed in algebraic manipulation (teachers among them), it may soon become totally imperceptible. (pp.198-199)

The duality (procedure vs. structure) of algebraic expression results in student learning difficulties. For example, when knowing x=3, y=2, find out the value of 3x+y, a procedure perspective is adopted. While simplifying the expression of 3x+y+8x, it is necessary to adopt an object (structure) perspective. In addition, it was found that extrapolating some manipulation rules to other contexts inappropriately is a common mistake (Matz, 1982). For example, applying the distribution law o(AnB)=oAnoB to an inappropriate situation: $\sin(\alpha + \beta) = \sin(\alpha) + \sin(\beta)$ or applying " $\frac{AX}{A} = X$ " to " $\frac{AX + BY}{X + Y} = A + B$ " are common mistakes. For another example, applying a known rule (such as if ab=0 then a=0 or b=0) to an unfamiliar situation: (X-A)(X-B)=K $\Rightarrow$ (X-A)=K or (X-B)=K is also a common mistake.

Equation. An *equation* is a combination of letters, operations, and an equal sign such that if numbers are substituted for the letters, either a true or false proposition results. Aspects of student difficulty within this topic are well documented in the literature on students' algebra knowledge (Booth, 1984; Kieran, 1992; Wagner, Rachlin, & Jensen, 1984; Wagner & Kieran, 1989).

First of all, it is not easy to understand the meaning of the sign "=". In algebra, equal sign "=" means equivalence of two algebraic expressions (equation), or presents one expression by another one (computation). Alibali, Knuth, Hattikudur, Mcneil and Stephens (2007) investigated

81 students at grades 6, 7 and 8 about their understanding of equal sign and equation for three years. They found that overall the students increased their understanding of the two concepts, but some eighth students still did not understand the equal sign completely.

Second, when solving equations, students face two challenges: the meaning of equal sign and the reverse computation relationship between addition and subtraction (Booth, 1984; Sfard & Linchevski, 1994). For example, Sfard and Linchevski (1994) found that students at the ages of 14 and 15 were able to solve the equation 7x+157=248, but failed to solve the equation 112=12x+247. They attribute the students' difficulty to two issues: the position and meaning of the equal sign "=", and the subtraction of a larger number from a smaller number. Moreover, even though students understood the reverse computation relationship between addition and subtraction, they still did not understand that the order of computation cannot be changed arbitrarily. For them, it is still a big challenge to solve equations including combination computation (Piaget & Moreau, 2001; Ronbing, Ninowski, & Gray, 2006).

2.4.3 Two Perspectives about the Concept of Function: A Case Study of Quadratic Function

Process-product dichotomy is a widely accepted theory of mathematics concept development (Briedenbach et al., 1992; Schwartz & Yerushalmy, 1992; Sfard, 1992). With regard to the development of the function concept, according to the process perspective, a function is perceived as linking x and y values: for each value of x, the function has only one corresponding y value. On the other hand, the object perspective regards functions or relations and any of its representation as entities. For example, functions could be regarded algebraically as members of parameterized classes, or in the plane graphs could be thought of as being rotated or translated (Moschkovich, Schoenfeld, & Arcavi, 1993).

Researchers have further provided more detailed descriptions and illustrations of these two perspectives (Breidenback et al., 1992; Even, 1990; Moschkovick et al., 1993; Schwartz & Yerushalmy, 1992; Sfard, 1992). For example, Schwartz and Yerushalmy (1992) illustrated the process-product distinction as follows:

Consider the two functions: x+3 and 4+x-1.

From the point of view of the process that is carried out with the recipe, these are two different recipes. If, however, one was to plot the output of each of these recipes again its input on a Cartesian plane then the two recipes would be indistinguishable. We see that the symbolic representation of function makes its process nature salient, while the graphical representation suppresses the process nature of the function and thus helps to make the function more entity-like. A proper understanding of algebra requires that students be comfortable with both of these aspects of function (p.265).

Furthermore, Moschkovick et al. (1993) not only extensively illustrated the distinction between the process and object perspectives, but also emphasized the importance of the connection between these two perspectives, and the flexibility in switching from different perspectives. They discussed multiple methods of solving the following question:

Why is the graph of y=3x steeper than the graph of y=2x? What about y=4x, y=5x, y=10x?

In one method, by considering the equation of line L: $y = mx + b$ and two points $P(x_1, y_1)$ and $Q(x_2, y_2)$, an algebraic formula was resulted in: $m = \frac{y_2 - y_2}{x_1 - x_2}$. If taking any two points on L whose x coordinates differ by 1, then $y_2 = y_1 + m$. If m is positive, the graph of L raises m units for each unit change in x. Thus, large (positive) m corresponds to steeper slope. This interpretation   basically adopted a process perspective.

In another method, letting $L_1: y = m_1x$, and $L_2: y = m_2x$ pass respectively through the points (1, m_1), and (1, m_2), hence $m_2 > m_1$, L_2 rise more steeply. In this solution, two aspects of both the process and object perspective were adopted in the algebraic and graphical representations. From the object perspective, the individual equations and lines are considered as members of the parametric family $\{y = mx : m \in R\}$. But using points on the graphs and determining their coordinates using the equations of the lines, employs the process perspective.

2.4.4 Flexibility in Learning the Concept of Function: A Case Study of Quadratic Function.

In this section, I review the meanings of flexibility in using representations, and summarize the studies on teachers' knowledge of algebra for teaching with regard to the flexibility in using representations.

Flexibility in using representations. Learning algebra with understanding requires students to "understand the meaning of equivalent forms of expressions, equations, inequalities, and relations" (NCTM, 2000, p. 296). In order to create that understanding, teachers have to organize a classroom discussion to open questions about the equivalence. For example, with regard to quadratic equations or functions, students need opportunities to discuss questions across equations, expressions and functions as follows:

When solving $3x^2 + 3x + 3 = 0$, I can think of the task as finding the zeros of the function $y = 3x^2 + 3x + 3$. In the context of finding zeros, I can divide 3 in the equation. However, when working with the function of $y = 3x^2 + 3x + 3$, we cannot divide all coefficients by 3. Why is that? (Chazan & Yerushalemy, 2003, p.124)

In the *Focal Points from pre-K to Grade 8* (NCTM, 2006), students (grade 8) are suggested to use linear functions, linear equations, and systems of linear equations to represent, analyze, and solve a variety of problems. Students are expected to

1. Recognize a proportion (*y/x = k*, or *y = kx*) as a special case of a linear equation of the form *y = mx + b*, understanding that the constant of proportionality (*m*) is the slope and the resulting graph is a line through the origin.

2. Understand that the slope (*m*) of a line is a constant rate of change, so if the input, or *x*-coordinate, changes by a specific amount, *a*, the output, or *y*-coordinate, changes by the amount *ma*.

3. Translate among verbal, tabular, graphical, and algebraic representations of functions (recognizing that tabular and graphical representations are usually only partial representations).

4. Describe how such aspects of a function as slope and y-intercept appear in different representations. (NCTM, 2006, p. 20).

In the *Focus of High School Mathematics* (NCTM, 2009), sense making and reasoning is the core value of learning mathematics in general, and algebra in particular. It was suggested that key elements of reasoning and sense making with algebraic symbols should include the following:

1. *Meaningful use of symbols.* Choosing variables and constructing expressions and equations in context; interpreting the form of expressions and equations; manipulating expression so that interesting interpretations can be made.

2. *Mindful manipulation.* Connecting manipulation with the laws of arithmetic; anticipating the results of manipulations; choosing procedures purposefully in context; picturing calculations mentally.

3. *Reasoned solving.* Seeing solution steps as logical deductions about equality; interpreting solutions in context.

4. *Connecting algebra with geometry.* Representing geometric situations algebraically and algebraic situations geometrically; using connections in solving problems.

5. *Linking expressions and functions.* Using multiple algebraic representations to understand functions; working with function notation. (NCTM, 2009, p. 31)

In order to develop algebra fluency, attention should be paid to interpret expressions both at the formal level and as statements about real-world situations. At the outset, the reasons and justifications for forming and manipulating expressions should be a major emphasis of instruction (Kaput et al., 2008). As comfort with expressions grows, constructing and interpreting them require less and less effort and gradually become almost subconscious. For example, students should know which is most useful for finding the maximum value of the quadratic function:

$$1\frac{3}{16} + 18t - 16t^2, (t - \frac{19}{16})(t + \frac{1}{16}), (t - \frac{9}{16})^2 + \frac{100}{16}.$$

Although multiple representations of functions—symbolic, graphical, numerical, and verbal—are commonly seen, the idea of multiple *alge-*

braic representations of functions is not often made explicit. Different but equivalent ways of writing the same function can reveal different properties of the function (as illustrated by the above example). Building fluency in working with algebraic notation that is grounded in reasoning and sense making will ensure "that students can flexibly apply the powerful tools of algebra in a variety of contexts both within and outside mathematics" (NCTM, 2009, p. 37).

Function is one of the most important tools for helping students make sense of the world around them and prepare them for further study in mathematics as well. Students' continuing development of the concept of function must be rooted in reasoning, and conversely, functions serve as an important tool for reasoning. Key elements of reasoning and sense making with functions include the following (NCTM, 2009, p. 41):

1. *Using multiple representations of functions.* Representing functions in various ways, including tabular, graphic, symbolic (explicit and recursive), visual, and verbal; making decisions about which representations are most helpful in problem-solving circumstances; and moving flexibly among those representations.

2. *Modeling by using families of functions.* Working to develop a reasonable mathematical model for a particular contextual situation by applying knowledge of the characteristic behaviors of different families of functions.

3. *Analyzing the effects of parameters.* Using a general representation of a function in a given family (e.g., the vertex form of a quadratic, $f(x) = a(x - h)^2 + k$ to analyze the effects of varying coefficients or other parameters; converting between different forms of functions (e.g., the standard form of a quadratic and its factored form) according to the requirements of the problem-solving situation (e.g., finding the vertex of a quadratic or finding its zeros).

In summary, these documents suggested that the following aspects are important in algebra learning, particular with the learning of function: (1) building the connection among expressions, equations/inequality, and functions; (2) flexible use of multiple representations of a function and shift among different representations, and (3) flexible use of multiple expressions of a function. As Star and Rittle-Johnson (2009) argued, "understanding in algebra can be considered to consist of two complementary capacities, which we refer to as between and within representation fluency. The first concerns the ability to operate fluently between and across multiple representations, while the second is about facility within each individual representation" (p. 11). Thus, it is critical to have a flexible and adaptive use of representations and expressions.

The representation flexibility should include the following abilities: (1) Having the necessary diagrammatic knowledge to interact with the representations (de Jong et al. 1998; Roth & Bowen, 2001); (2) Being able to coordinate the translation and switching between representations within the same domain (de Jong et al. 1998; Gagatsis & Shiakalli, 2004; Lesh, Post, & Behr 1987); and (3) Having the necessary strategic knowledge and skills to choose the most appropriate representation for each occasion (Uesaka & Manalo, 2006).

With regard to the concept of functions, it is necessary to consider flexibility in two aspects. One is the flexibility in selecting perspectives of function: process and object, and shifting between these two perspectives (Breidenbach et al., 1992; Dubinsky & Harel, 1992; Sfard, 1991). Another is the flexibility in using appropriate representations of functions: tabular, graphic, symbolic, and verbal representations, and shifts between them (Even, 1998; Moschkovick et al., 1993).

Teachers' knowledge of representational flexibility. Studies have found that teachers do not have the appropriate knowledge of using representation flexibly (Black, 2007; Even, 1998). To investigate prospective mathematics teachers' subject knowledge, Even (1998) examined teachers' knowledge of using representations. For example, she presented teachers with the following question:

If you substitute 1 for x in expression $ax^2 + bx + c$ (a, b and c are real numbers), you get a positive number, while substituting 6 gives a negative number. How many real solutions does the equation $ax^2 + bx + c = 0$ have? Explain.

Only 14% of the 152 subjects correctly solved the problem. These subjects considered the function corresponding to $y = ax^2 + bx + c$, switched representations, and either referred to a graph mentally or actually sketched a graph. Most of the subjects (about 80%) did not show any attempt to look at another representation of the problem, and did not solve the problem. A large number of subjects were stuck in the manipulation of inequalities: $a + b + c > 0$, $36a + 6b + c < 0$. As a result, they were not able to reach correct solutions.

In addition, she also found that the subjects who used a point-wise approach (e.g., process perspective) were more successful in solving problems that involved different representations of function than subjects who used a global approach (e.g., object). For example, in the following question:

This is the graph (Figure 2.2) of the function $f(x) = ax^2 + bx + c$. State whether a, b, and c are positive, negative or zero. Explain your decision.

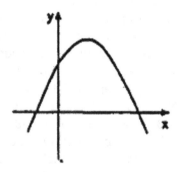

Figure 2.2. Graph of quadratic function

Only subjects who used a point-wise approach and looked at the y-intercept found correctly the sign of "c". They explained as follows: c is positive because when x=0, f(x)=c is positive).

In Black's (2007) study, he asked in-service high school mathematics teachers to explain their choice to the following question to students. 15 of 67 participants arrived at the correct answer.

Mr. Seng's algebra class is studying the graph of $y = ax^2 + bx + c$ and how changing the parameters of a, b, and c will cause different translations of the original graph (Figure 2.3).

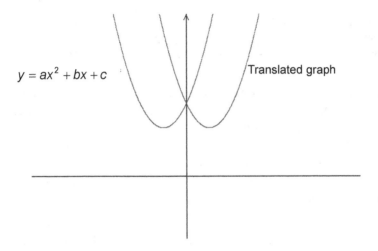

$y = ax^2 + bx + c$

Translated graph

Figure 2.3. Graphs of original and translated quadratic functions

Which of the following is an appropriate explanation of the translation of the original graph $y = ax^2 + bx + c$ to the translated graph?

A. Only the **a** value changed.
B. Only the **c** value changed.
C. Only the **b** value changed.
D. At least two of the parameters hanged.
E. You cannot generate the translated graph by changing any of the parameters.

He concluded that the participants had "a lot of difficulty in both their answer selection, as well as in the explanations provided for those answers" (Black, 2007, p. 134). Additionally he found that 13 of the 67 participants did not even attempt to answer the problem, and many more did not explain their answers. The finding seemed to suggest those mathematics teachers had difficulty in performing on functions as an *objects* (i.e., translation of graph) and shift between different representations (Using algebraic representation to explain graphic changes).

As argued by Moschkovick et al. (1993),

> "Competence in the domain (i.e., line equation) consists of being able to move flexibly across representations and perspectives, where warranted: to be able to 'see' lines in the plane, in their algebraic form, or in tabular form, as objects when any of those perspective is useful, but also to switch to the process perspective, where that perspective is appropriate" (p. 97).

Given the importance of developing students' flexibility in learning the concept of function, and the weakness of teachers' knowledge for teaching functions that promote this flexibility, I focus on teachers' flexibility in shifting between different perspectives (process vs. object), and selecting multiple representations (verbal, tabular, symbolic, and graphic). Further, I will illustrate the concept of flexibility in solving problems related to quadratic function in the methodology section.

2.5   Mathematics Teacher Education Systems in China and the U.S.

In this section, I briefly introduce teacher preparation systems in China and the U.S. to provide a background for understanding teacher knowledge growth.

2.5.1 Mathematics Teacher Education in China

Since adopting the "nine-year compulsory education system" in 1986, teacher education has become a daunting task. Through approximately 20 years of effort, a three-stage "normal" education has been established and has made significant contribution to educating teachers from elementary to secondary levels in China. It means: (1) primary school teachers are trained in secondary normal schools; (2) junior high school teachers are trained in three-year teacher colleges; and (3) senior high school teachers are trained in four-year teacher colleges and normal universities. However, with the rapid development of the economy and technology in China, it has been an urgent agenda to upgrade and foster teachers' quality. In order to meet this challenge, the Ministry of Education (1998) documented an action plan for revitalizing education in the 21st century. Two projects were launched; one is referred to as the "Gardener Project" and aims to establish continuing teacher education systems for practicing teachers. Meanwhile, the Ministry of Education (1999) enacted *a decision on deepening education reform and whole advancing quality education* in which comprehensive universities and non-normal Universities were encouraged to engage in educating elementary and secondary teachers. This meant that the privilege of normal universities for teacher education changed. The Ministry of Education (2001a) put forward a process to improve an open teacher education system based on the existing Normal universities and supported by other universities, and to integrate prospective teacher preparation and practicing teachers' professional development.

Through five years of development and research, teacher education has seen some changes:

1. Integration of education of prospective and practicing teachers;

2. Opening of teacher education in all qualified universities rather than just Normal universities;

3. Forming a new three-staged teacher education: where primary school teachers are trained in the three-year teacher colleges or four-year teacher colleges; the junior and senior high school teachers are trained in four-year teacher colleges and Normal universities, and some of the senior high school teachers are required to attain postgraduate level studies (Gu, 2006).

In 2004, there were more than 400 institutes conducting teacher education programs and about 280 of them were teacher education universities or colleges. It was also found that one third of graduates who became teachers were from non-teacher education institutes (Yuan, 2004). This proportion has steadily increased in recent years. According to the *Educational Statistics Yearbook of China 2008* (Ministry of Education, 2009), there are only 188 normal institutions where preparing teachers at different levels is the main purpose. In addition, there is now a flexible and encouraging accreditation system for teacher recruitment in China. University degree holders who wish to become secondary school teachers can pass the related examinations, usually pedagogy, psychology, and subject didactics.

Middle and high school mathematics teacher preparation programs were typically hosted in mathematics departments. Through analysis of the course design in a mathematics department at a normal university (Li, Huang, & Shin, 2008), it was found that prospective mathematics teachers were required to complete 63 credit hours in mathematics content, 6 credit hours in pedagogical courses, and 14 credit hours in student teaching and thesis. In addition, there were 20 credit hours in elective mathematics and mathematics education courses. Li and colleagues (2008) concluded that the secondary mathematics preparation program exhibits the following characteristics:

1. Providing prospective teachers with a strong foundation in profound mathematics knowledge and advanced mathematics literacy

2. Emphasizing review and study of elementary mathematics. It was believed that a profound understanding of elementary mathematics and strong ability of solving problems in elementary mathematics were crucial to being a qualified mathematics teacher at secondary schools. Due to the tradition of examination oriented teaching, a high level of problem solving ability is necessary for a qualified teacher;

3. Teaching practicum is limited. A six-week teaching practicum can only provide prospective teachers with a preliminary experience of teaching in secondary schools. (Li et al., 2008, p. 70)

This reflects the belief that a solid mathematics base is vital for teacher preparation. Furthermore, taking more advanced mathematics courses is on a higher priority since prospective teachers will not have many opportunities to learn them in their career. The main goal is to foster prospective teachers with a bird's eye view of understanding elementary mathematics deeply, even though there are special courses such as Modern Mathematics and School Mathematics, which connect advanced mathematics to elementary mathematics. In contrast with the high requirement in mathematics content, teacher educators normally believe that prospective teachers can develop pedagogical skills and knowledge later during their teaching.

2.5.2 Mathematics Teacher Education in the U.S.

In the United States, there are three levels of pre-college education: elementary school (Grades K-5 or K-6); middle school or junior high school (Grades 6-8 or 7-8); and high school (Grades 9-12). In 2007, forty states had either a middle school or junior high school certification or endorsement requirement (National Middle School Association, 2007). Many of these states have special mathematics requirements for that certification or endorsements by the teachers' selected area of content expertise. In mathematics, these special requirements range from passing a test to completing the equivalent of an undergraduate minor in mathematics. The Conference Board of Mathematical Science [CBMS] (2001) recommendations call for the teaching of mathematics in middle school (Grades 5-8) to be conducted by mathematics specialists; teachers specially educated to teach mathematics to the students of the grade levels they instruct. These teachers should have at least twenty-one semester hours in mathematics, including at least twelve semester hours of fundamental ideas of mathematics appropriate for middle school teachers.

Middle school math teachers. Based on National NAEP survey (Smith, Arbaugh, & Fi, 2007), 85% of the nation's eighth graders are taught by teachers who were certified by their state. When examined by teachers' degrees, 30% of the nation's eighth graders had teachers with an undergraduate degree in mathematics; 26 % had teachers with an undergraduate degree in mathematics education; and the remaining students were taught by a teacher with a degree in some other discipline. Thus, at least one-third of the nation's eighth-grade students were being taught mathematics by teachers without substantial mathematics training. According to National Report Card 2009 (National Assessment of Education Progress [NAEP], 2009), the situation gets worse: 27% of the nation's eighth graders had teachers with an undergraduate degree in mathematics; 30 % had teachers with an undergraduate degree in mathematics education. This is a major concern of U.S. mathematics teacher education.

High school mathematics teachers. For high school mathematics teacher certification, states require from eighteen to forty-five semester hours of mathematics, equivalent to six to fifteen semester courses, or they require a major in the subject. When a number of credit hours of mathematics are specified for the certificate, almost half require thirty credit hours. The specific courses mentioned include three courses in calculus (two single-variable and one multivariable), linear algebra, geometry, and abstract algebra, plus a number of electives.

The 2000 National Survey of Science and Mathematics Education (Whittington, 2002) was designed to identify trends in the areas of teacher background and experience, curriculum and instruction, and the availability and use of instructional resources. A total of 5,728 science and mathematics teachers in schools (1,367 of them were high school mathematics teachers) across the United States participated in this survey. According to this survey, 58% of mathematics teachers in grades 9-12 in their sample had an undergraduate major in mathematics, 22% had a degree in mathematics education, 10% had a degree in some other education field, and 10% had a degree in a field other than education or mathematics. In this sample, 96% of teachers had completed a course in calculus, 86% in probability and statistics, 83% in geometry, 82 % in linear algebra, 70% in advanced calculus, and 65% in differential equations and so on. See table 2.1.

Table 2.1. High School Mathematics Teachers Completing Various College Courses

Course	Percent of teachers
General methods of teaching	90
Methods of teaching mathematics	77
Supervised student teaching in mathematics	70
Instructional uses of computers/other technologies	43
Mathematics for middle school teachers	26
Geometry for elementary/middle school teachers	17
Calculus	96

Course	Percent of teachers
Probability and statistics	86
Geometry	83
Linear algebra	82
College algebra/trigonometry/ elementary functions	80
Advanced calculus	70
Computer science course	68
Differential equations	65
Abstract algebra	65
Computer programming	62
Other upper division mathematics	60
Number theory	56
History of mathematics	41
Real analysis	38
Discrete mathematics	38
Applications of mathematics/ problem solving	37

The CBMS report (2001) recommends that high school teachers of mathematics have a major in mathematics; that includes a six-hour capstone course connecting their college mathematics course with high school mathematics. This recommendation stems from the view that teachers need to know the subject they will teach, they need to understand the broad range of the mathematical sciences their students will encounter in their careers, and they need to develop the habits of mind and dispositions towards doing mathematics that characterize effective workers in the field.

2.6 Comparative Studies on Teachers' Knowledge for Teaching between China and the U.S.

Ma's (1999) revealed Chinese elementary teachers had a profound understanding of fundamental mathematics concepts in four areas: subtraction with regrouping, multi-digit multiplication, division by fractions, and the relationship between perimeter and area, in comparison with U.S. counterparts. After that, several studies have focused on mathematics teacher knowledge in China and the U.S. (An et al., 2004; Cai, 2005; Cai & Wang, 2006; She, Lan, & Wilhelm, 2011; Zhou, Peverly, & Xin, 2006). By comparing pedagogical content knowledge of middle school mathematics teachers between the U.S. and China, An et al. (2004) found that the Chinese mathematics teachers emphasized gaining correct conceptual knowledge by relying on a more rigid development of procedures, while the United States teachers emphasized a variety of activities designed to promote creativity and inquiry in order to develop concept mastery.

A study investigating Chinese and US middle school teachers' ways of solving algebraic problems (She, Lan, & Wilhlem, 2011) revealed that U.S. teachers were more likely to use concrete models and practical approaches in problem solving, but they seemed to lack a deep understanding of underlying mathematical theories. The Chinese teachers were inclined to utilize general roles/strategies and standard procedures for teaching, and they demonstrated an interconnected knowledge network when solving problems.

In addition, a study comparing 162 U.S. and Chinese third grade mathematics teachers' expertise in teaching fractions (Zhou et al., 2006) found that Chinese teachers significantly outperformed their U.S. counterparts in subject matter knowledge, but they performed poorly in comparison to their U.S. counterparts on a test designed to measure general pedagogical knowledge. However, in pedagogical content knowledge there are no determinative patterns found.

Cai and his colleagues (Cai, 2000, 2005; Cai & Wang, 2006) have conducted a series of comparative studies on students' problem solving and teachers' construction of representations between China and the U.S. It was found that Chinese students preferred using symbolic representations while U.S. students tended to use pictorial representations (Cai, 1995, 2000). In addition, both Chinese and U.S. teachers used concrete representations for developing the concepts of ratio and average, but Chinese teachers tended to use symbolic representations for solving problems while U.S. teachers still preferred to use concrete representations (Cai, 2005; Cai & Wang, 2006). Furthermore, Huang and

Cai (2011) found that the U.S teachers tended to develop multiple representations simultaneously while the Chinese teachers tend to selectively use representations hierarchically.

These studies seem to suggest Chinese elementary mathematics teachers have a stronger subject matter knowledge, probably pedagogical content knowledge, compared with their U.S. counterparts. Meanwhile, Chinese mathematics teachers value symbolic representation more than U.S. mathematics teachers when solving problems.

However, there are no comparative studies on teachers' knowledge needed for teaching special areas at middle and high school levels between China and the U.S. Also, teachers' representational flexibility, which is closely related to teachers' beliefs and teaching has not been explored appropriately. Thus, in this study, I examined prospective secondary school teachers' knowledge of algebra for teaching with a focus on the concept of function and representation flexibility in China and the U.S. Therefore, the current study will contribute to the understanding of mathematics teacher's knowledge of algebra for teaching in China and U.S. at middle and high school levels in addition to shedding light on improvement of teacher preparation programs in China and the U.S.

2.7 Conclusion

This chapter provided a review of relevant literature, laying a theoretical foundation for this study. First, the development of the notion of teacher knowledge in general, and ways of defining and measuring mathematics teacher knowledge needed for teaching in particular were summarized. Second, the specific frameworks for studying mathematics knowledge of algebra for teaching were analyzed and compared. Third, relevant studies on algebra teaching and learning, and teacher knowledge needed for teaching algebra were summarized. Fourth, the literature review focused on teacher knowledge for teaching the concept of function promoting flexibility in adapting appropriate perspectives and representations of function. Fifth, a brief summary of mathematics teacher preparation in China and the U.S. was presented. Finally, comparative studies on teachers' knowledge for teaching between China and the U.S. were summarized.

3 Chapter Three: Methodology

This study compared the characteristics of mathematics knowledge needed for teaching algebra between China and the United States. An embedded mixed methods design was adapted (Creswell & Clark, 2007). The design of the main data set consists of written answers to a questionnaire which includes multiple choice items and open-ended items, while the supportive data set is comprised of the written answers to the open-ended items and follow up interviews. The primary purpose of this study was to compare the status and structure of teacher knowledge of algebra for teaching through quantitatively analyzing the participants' performance in the KAT survey between the two countries. The second purpose was to further illustrate the similarities and differences in KAT through qualitatively analyzing the answers to the open-ended questions and follow up interviews, which focus on the core concept of function. Based on the questionnaire of KAT by Floden and McCrory (2007), I developed an instrument for measuring teachers' knowledge needed for teaching algebra, with a focus on the concept of function. Then, completed questionnaires were collected from 376 prospective Chinese mathematics teachers from five purposefully selected teachers' training institutions. At the same time, I also collected data from 115 U.S. prospective teachers who were preparing to be mathematics and science teachers at the middle school level from a well respected public university in the southern United States. All of the Chinese and U.S. data were scored and quantified as an SPPS data set, and then analyzed through techniques including multiple comparisons and SEM model to answer the research questions. A qualitative analysis of the answers to the open-ended items was conducted to identify the strategies used and the flexibility in solving these problems. This chapter is organized into three parts. First, the process of developing a Chinese instrument based on the existing U.S. questionnaire is described in detail. Next, the ways of data collection are described. Finally, the methods of data analysis are depicted.

3.1   Instrumentation

The translation equivalence and cultural adaptation of instruments in international comparative studies is an issue that has to be dealt with carefully and appropriately. For example, the TIMSS technical report (Chrostowski & Malak, 2004) suggested that translators should consider the following aspects when translating instruments: (1) identifying and minimizing cultural differences, (2) finding equivalent words and phrases, (3) ensuring that the reading level was the same in the target language as in the original international version, (4) ensuring that the essential meaning of the text did not change, (5) ensuring that the difficulty level of achievement items did not change, and (6) making changes in the instrument layout required due to translation. Both TIMSS and PISA (Organization for Economic Co-operation and Development [OECD], 2006) suggested adopting a double translation procedure (*i.e.* two independent translations from the source language, with reconciliation by a third person). This strategy offers two significant advantages when compared with the back translation procedure: (1) Equivalence of the source and target languages is obtained by using three different people (two translators and a reconciler) who all work on the source and the target versions; and (2) Discrepancies are recorded directly in the target language (OECD, 2006).

With regard to the adaptation of instruments of mathematics knowledge for teaching to different cultures, Delaney et al. (2008) highlighted several critical issues. These points include: (1) what teachers do during mathematics lessons, (2) teachers' conceptions about mathematics and about mathematics teaching, (3) the classroom contexts in which the knowledge is used, (4) differences in the types and sophistication of the explanations of students mistakes, (5) responding to student errors, (6) the presence and prevalence of specific mathematical topics, (7) the mathematical language used in the school, and (8) the content of the textbooks. They further suggested making relevant changes in the following four categories: the general cultural context related, the school cultural context related, mathematical substance related, and others.

In this study, I dealt with these issues using the following strategies: (1) content appropriateness in different cultures, and (2) equivalence of the instrument translation.

3.1.1 Content Appropriateness

A research team at Michigan State University developed an instrument for measuring mathematics knowledge of algebra for teaching. The validity and reliability of the instrument has been tested in the United States context (Floden et al., 2009). In order to adapt this U.S. instrument into the Chinese context, I first translated it into Chinese and invited three Chinese mathematicians and mathematics educators who are bilingual scholars to scrutinize the instrument and provide supplementary items if needed. Since the function concept is a core concept in algebra and is difficult for students to learn and for teachers to teach (Kieran, 2007; Even, 1990, 1993), I decided to include some open-ended items related to teacher knowledge needed for teaching the concept. Based on (1) the study of the mathematics curriculum standards in China and the U.S., (2) an extensive literature review on mathematics knowledge for teaching the concept of function, and (3) suggestions made by Chinese mathematicians and mathematics educators, five open-ended questions were adapted or designed. These open-ended items were reviewed by two mathematics educators and one mathematician at the sample U.S. University. One of the mathematics educators had extensive teaching experience at secondary schools (including middle and high schools) and universities in the U.S. The other mathematics educator had secondary mathematics teaching experience in China and several years of experience as a faculty member in math education programs in China and the U.S. These items were determined appropriately for secondary (i.e., middle and high school) mathematics teachers. Both examined these items carefully and improved the wording of some items. The mathematician also gave very detailed comments and corrections. Based on the feedback from them, I made a final version in English.

3.1.2 Translation Equivalence

The English instrument was translated into Chinese by the researcher and another Ph.D. candidate in mathematics education from China. A third bilingual mathematics educator at the sample university in the U.S. compared these two Chinese versions and made a final version through discussions with the researcher.

3.1.3 Appropriateness of the Survey from Teachers' Perspectives

In order to understand teachers' perceptions of the questionnaire, we invited ten prospective teachers in the last year of their four year bachelor degree program in secondary mathematics education from an eastern city of China and four in-service teachers from a secondary school in a northern city of China (two middle school teachers, and two high school teachers) to complete this survey within 45 minutes. Based on their written answer sheets, I identified several items that were commonly missed so that they could be used during the interview process.

I interviewed three of the four in-service teachers who completed this survey through a videoconference system. One of them is a high school teacher with a senior position (Equivalent to an associate professor at universities in the U.S.). The others are middle school mathematics teachers. Both of them had more than six years of teaching experience. The interview was aimed at understanding participants' opinions about the overall difficulty, the appropriateness of items, and their interpretation of the identified mistakes.

These teachers expressed the following opinions in the interviews. First, they perceived the questionnaire to be relatively easy, but the items were flexible, contextual, and covered broad topics. Second, they felt that the questions related to the advanced mathematics topics were the most difficult. Third, in general, the expression and background of problems were clear and easy to understand, but some of the problems were so novel that they had to be cautious in order to ensure that they correctly understood them. Fourth, the questionnaire was a little long to complete within 45 minutes. In addition, they believed the questionnaire could measure a teachers' knowledge of algebra for teaching. Finally, they provided some suggestions for improving the questionnaire, which included adding an introduction to the purpose of the questionnaire and refining the format of the questionnaire.

Based on the interview results, I have made improvement on in the Chinese questionnaire which includes 25 items. 20 items, including 17 multiple choice items and 3 open ended items, were translated from the original English questionnaire, and an extra 5 open-ended items were created by the researcher. The format of the questionnaire was compacted from the original 14 pages to 8 pages. This includes the introductory information and 25 items. In addition, some specific terms, which are easily misunderstood, were highlighted with bold font. For instance, in Item 11, large and bold words were used to highlight to select the "NOT appropriate" one.

Finally, a questionnaire booklet was developed in this study. It includes two parts. The first part includes 17 multiple-choice items and 8 open-ended items (items 18 to 25) with a focus on teachers' knowledge for teaching the concept of function. The second part is an answer booklet, including participants' backgrounds such as current grade and courses taking, a multiple choice answer table, and open-ended answer sheets.

Three U.S. students who were studying in a Ph.D program in math education were invited to complete this survey within 45 minutes. This small pilot showed that the time for completing the survey was appropriate and participants were able to understand and answer each of the questions. Thus, no further revision of the questionnaire appeared necessary. In the next section, justification for the adaptation of the open-end question will be given.

3.1.4 Measuring Knowledge for Teaching the Concept of Function

Based on an extensive literature review of teachers' knowledge for teaching the concept of function, I focused on two aspects of the concept of function: fluency and flexibility of knowledge for teaching function in general and for teaching quadratic functions in particular. With regard to the first aspect, I focused on the understanding of the concept of function in terms of shifting between two perspectives (*process vs. object*) appropriately. In addition to an original item 18 that required a process perspective to effectively answer the questions, I created two items. One item (item 24) that required using an object perspective in order to effectively complete the proof, and another (item 25), in which it is necessary to adopt the connection of the two perspectives in order to appropriately answer the item.

In order to measure teachers' knowledge for teaching quadratic functions, I focused on the fluency and flexibility of using different representations. Four items are used to gauge teacher knowledge for teaching this particular content field. Item 19 from the original questionnaire is used to measure knowledge of solving quadratic inequalities by using algebraic and graphic representations. Item 21 is used to measure teacher knowledge of flexibly using algebraic and graphic representations and the connection of function, equation and inequality. Item 22 is used to measure the flexible use of multiple representations of quadratic functions (graphic and algebraic) and translations of graphs. Item 23 is designed to use multiple forms of algebraic representations and translations between different representations. These items are shown as follows:

18. a) On a test a student marked both of the following as non-functions

(i) $f: R \to R$, $f(x) = 4$, where R is the set of all the real numbers.

(ii) $g(x) = x$ if x is a rational number, and $g(x) = 0$ if x is an irrational number.

For each of (i) and (ii) above, decide whether the relation is a function, and write your answer in the Answer Booklet.

b) If you think the student was wrong to mark (i) or (ii) as a non-function, decide what he or she might have been thinking that could cause the mistake(s).

Write your answer in the Answer Booklet.

(Adapted from Even (1993))

19. Solve the inequality $(x - 3)(x + 4) > 0$ in **two essentially** different ways. Show your work in the Answer Booklet.

21. If you substitute 1 for x in expression $ax^2 + bx + c$ (a, b and c are real numbers), you get a positive number, while substituting 6 gives a negative number. How many real solutions does the equation $ax^2 + bx + c = 0$ have?

One student gives the following answer:

According to the given conditions, we can obtain the following in-equations:

$a + b + c > 0$, and $36a + 6b + c < 0$.

Since it is impossible to find fixed values of a, b and c based on previous inequality, the original question is not solvable.

What do you think may be the reason for the students' answers? What are your suggestions to the student?

Write down your answers in as much detail as possible on your Answer Booklet.

(Adapted from Even (1998))

22. This item is adapted from Black (2007)) (See p. 27)

23. Given quadratic function $y = ax^2 + bx + c$ intersects x-axis at (-1, 0) and (3, 0), and its y-intercept is 6. Find the maximum of the quadratic function.

Show your work in as much detail as possible in the Answer Booklet.

(Adapted from NCTM (2009))

24. Prove the following statement:

If the graphs of linear functions $f(x) = ax + b$ and $g(x) = cx + d$ intersect at a point P on the x-axis, the graph of their sum function $(f + g)(x)$ must also go through P.

Show your work in the Answer Booklet.

25. When introducing the functions and the graphs in a middle school class (14-15 year-olds), tasks were used which consist of drawing graphs based on a set of pairs of numbers contextualized in a situation and from equations? One day, when starting the class, the following graph (Figure 3.1) was drawn on the blackboard and the pupils were asked to find a situation to which it might possibly correspond.

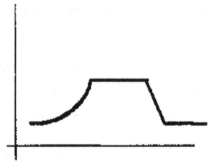

Figure 3.1. A graph presenting a daily life situation.

One student answered: 'it may be the path of an excursion during which we had to climb up a hillside, the walk along a flat stretch and then climb down a slope and finally go across another flat stretch before finishing.'

How could you answer this student's comments? What do you think may be the cause of this comment? Can you give any other explanations of this graph?

Write down your answers in as much detail as possible on your Answer Booklet.

(Adapted from Llinares (2000))

In summary, all the items and corresponding content areas, and types of knowledge are listed in Table 3.1.

Table 3.1. Items and Corresponding Content Areas and Knowledge Types

Item	Content area	Knowledge type
1	Expression	School mathematics
2	Equation	Teaching
3	Function	School mathematics
4	Function	Advanced mathematics
5	Equation	Teaching
6	Equation and expression	School Math
7	Function	Teaching
8	Function	Advanced math
9	Equation and expression	Advanced math
10	Function	Teaching
11	Function	Teaching
12	Equation and expression	Advanced math
13	Equation and expression	Advanced math
14	Equation	School math
15	Equation	Teaching
16	Function	Advanced math
17	Function	School math
18	Function	Teaching
19	Function	School math
20	Function	Advanced math
21	Equation and expression	Teaching
22	Function	Teaching
23	Function and equation	School math
24	Function and equation	Advanced math
25	Function	Teaching

3.2   Data Collection

In the following sections, I describe the process of data collection in China and the US.

3.2.1 Chinese Data Collection

In China, there is only one type of secondary mathematics teacher preparation programs (including middle and high school teachers). There are not specific programs for preparing middle school math teachers. These programs either are provided by normal universities or comprehensive universities. Usually, these programs are housed in a mathematics department (Li et al., 2008).

First, I discussed the selection of representative universities in China with a professor in a leading teacher education university in China. This professor was the former president of the National Higher Teacher Education University Association. I raised two criteria: (1) the score of university entrance examination, and (2) a wide spectrum of different programs.   With the help of the professor (and my personal connections with teacher education universities), I contacted math educators from seven Universities (Based on the educational institution list league in China in 2009, two belong to type 1 (top 10), two belong to type 2 (top 20), and three belong to type 3 (after 30)) through e-mails and phone contact. I explained the research purposes and the requirement of administering the survey. First, participating students were required to have 45 minutes to complete a questionnaire. Second, ideally, about 60 junior and 60 senior students would need to be recruited to complete the survey. All target coordinators promised to help collect the data from their respective universities. I sent the questionnaire to respective coordinators from the seven universities (with written instructions of conducting the survey) in early spring 2010. I asked them to explain the survey to their students as part of their course work and requested that the students complete the survey seriously and honestly within 45 minutes.

Due to some physical difficulties (for example, in some universities, either the senior students or the junior students were in the process of teaching practicum), not all of them met the deadline and requirement.

Two universities were not able to collect their data in time (one is type 1 and the other is type 3). One university only collected junior students' data while another university only collected senior students' data. All of the completed questionnaires were sent to a professor in Shanghai, and the professor helped to scan the completed questionnaires into PDF files and e-mail them to me. I printed the questionnaires in March 2010. Finally, 376 completed questionnaires were used for data analysis after excluding 8 copies from universities 3 and 4, and 3 copies from university 5 due to lack of background information. The distribution of the competed questions is shown in Table 3.2.

Table 3.2. Demographic Information of the Chinese participants

University	Code	Junior	Senior
Type 1 University	1	59	50
Type 2 University	2	71	
Type 3 University	3	33	15
Type 3 University	4	48	52
Type 2 University	5		48

Note. According to the teacher education institution ranking list in China in 2009, the researchers split it up to three types: type 1 (top 10) is the highest, followed by type 2 (around top 20), and type 3 (after top 30).

3.2.2 U.S. Data Collection

In the U.S., high school mathematics teachers usually need to earn a bachelor's degree in math, with certain required credits in math education. However, there are different routes for training middle school math teachers. The first one is preparing middle school teachers as a part of the preparation of secondary school teachers. The second one is specifically preparing middle school teachers (i.e., math and science interdisciplinary approach). The third is preparing middle school teachers as an extension of the preparation of elementary teachers (Dossey, Halvorsen, & McCrone, 2008; Schmidt et al., 2007).

I contacted three instructors who taught the mathematics education courses for junior and senior students at a large public school in the Southern United States. I explained the research project and requested their assistance to administer my survey during part of their class duration (around 45 minutes). All of them allowed me to conduct the survey using their class.

Students were told that their participation in the survey is fully voluntary. Instructors introduced me to their students and allowed me to briefly introduce my research project. After describing my research purpose and appreciating students' participation in their survey, I delivered the questionnaires to students and then collected the completed questionnaires after 45 minutes. All of the students who attended those classes completed the questionnaires. In all, I collected 115 copies of questionnaires from the three classes. The demographic information of the U.S. sample is displayed in Table 3.3.

Table 3.3. Demographic Information of the U.S. Sample

Program	Soph.	Junior	Senior	Total
Gr.4-8	11	48	31	90
Gr.6-12	0	5	4	9
Other	3	11	2	16
Total	14	64	37	115

Note. Gr. 4-8 presents the program prepared for science and mathematics teachers from grades 4 to 8. Gr.6-12 presents the program prepared for science and mathematics teachers from grades 6 to 12. Other refers to certificates for teaching math and science at middle school.

The table illustrates that the majority (79%) of the participants registered in the interdisciplinary program of math and science teachers at grades 4 to 8, while very few (7%) studied for programs of math and science teachers at grades 6 to 12. The remaining small part (14%) just took some courses for a certificate in teaching math or science at the middle school level. Among the participants, the majority (87%) were junior and senior students; only a small proportion were sophomore students.

3.2.3 Interview of the Selected U.S. Participants

The research design was to interview a sample of participants from China and the U.S. to clarify participants' answers and probe their thoughts. In addition to their free responses, the interviewees were intended to also be probed uniformly and non-uniformly. The uniform probes were presented to all subjects and were based on the analysis of the pilot study and corresponding survey with in-service teachers. These probes represented themes that appeared in many of the written answers (such as mistakes in solving inequalities, and mistakes in explaining graphs). The non-uniform probing was based on the specific answers each subject gave to the questionnaire and was meant to clarify ambiguous answers and discover specific dimensions that seemed important.

However, based on a preliminary analysis of completed questionnaires from the U.S. and China, I found that Chinese participants provided very detailed and rich responses to the open-ended questions for our analysis of their thoughts and strategies. On the other hand, the U.S. participants provided relatively short and simple responses to the open-ended questions. Therefore, I decided to only conduct interviews with purposely-selected U.S. participants.

Based on a detailed analysis of the answers of the participants from a class, I identified eight potential interviewees in terms of their performance such as typical correct answers and mistakes. Five of them agreed to attend an interview. The interview was conducted individually during the week after completing the survey. Each interview lasted about 20 minutes, and was audio recorded.

These interviewees were studying in the interdisciplinary program of middle school mathematics and science teacher preparation. Information was also collected concerning the high school mathematics courses taken (5 courses include: (1) Algebra I, (2)Algebra II, (3) Geometry, (4)Pre-Calculus, and (5) Calculus) and college mathematics and mathematics education courses taken. One took only four high school mathematics courses, while the others took five. The following 17 college courses were included: (1) Structure of mathematics I, (2) Structure of mathematics II, (3) Basic concept of geometry, (4) Introduction to abstract mathematics, (5) Integration of mathematics and technology, (6)Problem solving in mathematics, (7) Integrated math , (8) Mathematics methods in middle school, (9)Student teaching, (10) Freshman mathematics laboratory, (11) Analytic geometry and calculus, (12) Calculus, (13) Foundation of discrete mathematics, (14)Several variable calculus, (15) Liner algebra I, (16) Differential equations , and (17) Advanced calculus I. The courses taken by each interviewee are summarized in Table 3.4.

Table 3.4. The Course Taking Situation of the Interviewees

Name	High school course taking	College courses taking	College courses being taken in the semester
Larry	(1)-(5)	(1)-(3),(6),(11),(12)	(4)-(7)
Jenny	(1)-(5)	(1)-(3),(6)	(4),(5),(7)
Kerri	(1)-(4)	(1)-(3),(6),(10)-(12)	(4),(5),(7)
Alisa	(1)-(5)	(1)-(6)	(7)
Stacy	(1)-(5)	(1)-(4),(6),(10)-(12)	(5),(7)

All interviewees had taken at least four mathematics courses in high school and averaged nine courses in college.

I designed some particular questions for each of the open-ended items. For example, on Item 18, I designed the following prompt questions: (1) How do you judge whether a relationship is a function or not? (2) What is the vertical line test? (3) What would you teach to your students? Can you give me an example?

For Item 19, I found some common mistakes, then I asked some general questions such as (1) When reading solving the inequality, what knowledge, skills and methods come to your mind? (2) What is your understanding about two essentially different way? (Algebraic or Geometric methods?) (3) What do you think about the following operations?

a. $(x-3)(x+4)>0 \rightarrow x-3>0$, $x+4>0$, then $x>3$ and $x>-4$.

b. $x^2 + x - 12 > 0$; $x(x + 1) > 12$; $x>12$, $x+1>12$;

c. $x^2 + x > -12$; $x^2 > x - 12$; $x > \sqrt{x - 12}$;

In addition, the participants were also probed whether they could recall graphing methods for solving quadratic equations or inequalities.

For Item 20, I asked the following questions: Someone answers "Yes" and gives proof as follows:

$$\text{When } A = \begin{bmatrix} p & q \\ r & s \end{bmatrix} = \begin{bmatrix} 0 & 0 \\ 0 & 0 \end{bmatrix}, \ B = \begin{bmatrix} t & u \\ v & w \end{bmatrix} \text{ then,}$$

$$A\Delta B = \begin{bmatrix} 0t & 0u \\ 0v & 0w \end{bmatrix} = \begin{bmatrix} 0 & 0 \\ 0 & 0 \end{bmatrix}.$$

What do you think about this "proof?"

For Item 21, I asked the following questions: (1) What are the reasons for the student to make his/her judgment? and (2) To find solutions, are there other methods you can suggest to the student?

The prompt questions for Item 22 included: (1) What are the effects of change of parameters of a, b, and c on the graph? (2) What algebraic manipulations may help you identify the key parameter(s)?

The prompt questions for Item 23 included (1) Which formula of quadratic function did you choose for finding the function? And (2) How can you find the maximum of a given quadratic function?

The prompt questions for question 24 included: (1) What does it mean by intersecting at a point on the x-axis? (2) What is the meaning of (f+g)(x)?

The probing questions for Item 25 were: (1) What are the missing parts of students' comment (two variables, X vs. Y)? and (2) How can you explain other real life situations by using this graph?

3.3   Data Analysis

The data analysis includes three phases: (1) quantifying the data: developing five level rubrics for quantifying the open-ended items; (2) analyzing KAT at item and structure levels; and (3) analyzing open-ended item qualitatively: focusing on problem solving methods or mistakes, and flexibility of using representations.

3.3.1 Quantifying the Data

For each multiple-choice item (items 1 to 17), the correct choice was scored as 1, while the wrong choice was scored as 0. For each open-ended item, I developed a five level rubric for scoring the answers from 0 to 4. For items 18, 19, 20 and 24, I adapted the rubrics from the original rubrics developed by Michigan State University with some modifications (treating blank and missing answers as 0) and specifications (adding some details). For example, for item 18, I developed the following rubric:

Table 3.5. Rubric for Coding Item 18

Score	Description
0	Blank or total wrong answers in (i) and (ii)
1	(I): (a) answer (i) is function, (ii) is not or inverse
	(b) explanation is missing or wrong
	OR (II): (a) answer (i) and (ii) are not function, but (b) give some relevant explanations.
2	(I): (i) (a) is correct: (i) and (ii) are function.
	(b) without explanation or giving wrong explanation
	Or (II): (a) one of (i) and (ii) is function, (b) give an correct explanation
3	To give the answers with the following elements: (a) Point out (i) and (ii) are functions ;(b) The explanations do not relate to the key element (multiple-to-one or one –to- one), rather some superficial features such as: the function (i) with constant value, and the function (ii) is not continuous or expressed by two expressions or there are many holes.

Score	Description
4	To give answers with the following elements:
	(a) point out (i) and (ii) are functions;
	(b) point out that there is only one unique value corresponding to each value from domain value (such as one x value corresponds one y value, multiple x values correspond to one y value, but does not include one x value corresponds multiple y values). Or point out the use of the vertical line test.

Based on the study of existing rubrics, I developed general criteria for scoring all open-ended items as follows:

0- Blank or providing useless statements;

1- Providing several useful statements without a chain of reasoning for the correct answers;

2- Giving a correct answer but the explanations or procedures with major conceptual mistakes;

3- Giving a correct answer and appropriate explanations or procedures, with some minor mistakes;

4- Giving a correct answer with appropriate explanations and procedures.

Furthermore, based on the specification of each open-ended item, I further developed different rubrics for all of the open-ended items (See Appendix **A**). For example, the rubric for item 22 is described in Table 3.6.

Table 3.6. Rubric for Coding Item 22

Score	Examples
0	Blank or useless statements
1	Gives partial features of graph when changing a, b, or c.
2	Selects C or D and gives some explanations, with some serious mistakes, such as if a is changed then the graph is moved up or down.

Score	Examples
3	Gives answer C. However, reasons are not appropriately explained such as only mentioning the invariance of a or c.
4	Selects C and provides correct explanations such as:

4 (continued)

- Since change of a leads change of the openness, thus a is not changed; since y-intercept is not changed, so c is not changed. Thus, it is only possible to change b.

- The translated graph is the symmetrical graph of original graph with regard to y-axis. So, symmetrical line $x = -\dfrac{b}{2a}$ should be changed. However, the openness of the graph is not changed, so a should be invariant. Thus, only b is changed to –b.

- If f(x) and g(x) are symmetrical with regard to y-axis, then g(x)=f(-x), thus b is changed to –b.

3.3.2 Inter-Rater Reliability

Based on a preliminary examination of the open-items of 20 copies of U.S. questionnaires and 20 copies of Chinese questionnaires, I developed and tested the appropriateness of the rubrics. Then, I fully applied the finalized rubrics to the coding of the U.S. questionnaire.

After that, another secondary mathematics teacher and I scored the open-ended items of 109 copies of Chinese questionnaires from the type 1 institution separately. The inter-reliabilities of the items are 97% for item 18, 94% for item 19, 97% for item 20, 95% for item 21, 93% for item 22, 98 % for item 23, 86% for item 24, and 93 % for item 25. The disagreements were solved through discussions between raters and specifying the rubrics. The second mathematics teacher scored all of the remaining questionnaires. I double-checked the codes of U.S. questionnaires, and 100 copies from the remaining Chinese questionnaires. The agreement was higher than 95%, and I made relevant corrections.

3.3.3 Developing Categories of Different Strategies of Solving Open-ended Items

There are a total of eight open-ended items. One of them is related to metric and logical inference (Item 20); Three items are related to function concepts (Items 18, 24, 25), while four are related to quadratic functions /equations /inequalities (Items 19, 21, 22, 23). For the metric item, the analysis was focused on the logical equivalence and metric operations.

A two dimensional framework was developed (see Table 3.7) for analyzing knowledge for teaching function concept. One aspect presents the perspectives of function concept (process vs. object), and the other presents different representations (verbal, tabular, algebraic, and graphical).

Table 3.7. A Framework for Investigating Alternative Perspective of Function in Typical Representations

Perspective	Verbal	Tabular	Algebraic	Graphical
Process				
Object				

At the beginning, I tried to apply this two-dimensional framework to analyze all of the open-ended items. However, that attempt was found to be too complicated to implement. I then applied the dimension of perspective of function to analyze items 18, 24, and 25 and the dimension of representations to analyze items 19, 21, 22, and 23.

With regard to Items 18, 24, and 25, the analysis centered on the perspectives adopted. The categories and relevant explanations are shown in Table 3.8.

Table 3.8. Categories and Explanations with Regard to Function Concept in General

Item	Process	Object
18	Pointing out corresponding relationship between domain and range (one-to-one; multiple-to-one) [18P]	Point out the features of function expressions and graphs (a constant value or two expressions or one line, many holes /un-continuous curve) [18O]
24	Let f(x) and g(x) intersect at x-axis (p, 0), then, the following statements are true: (1) f(p) = 0 → ap + b = 0 → p = -b/a; (2) g(p) = 0 → cp + d = 0 → p = -d/c; (3) f(p) = g(p) → b/a = d/c → ad = bc; (4) f(p) = g(p) → ap + b = cp + d → p = -(b + d)/(a + c); According to (f+g)(p) = f(p) + g(p), and above statements, the student shows (f+g) (p) = 0.[24 P]	Let f(x) and g(x) intersect at x-axis (p, 0), then, f(p) = 0, g(p) = 0. So, (f+g)(p) = f(p) + g(p) = 0 + 0 = 0. Thus, (f+g) (p)=0. [24O]
25	It is necessary for students have a connection between two perspectives and a shift between graphical representation and verbal representation. The diagram could be interpreted as the following relations: (1) Height/distance vs. time [25C1]) (2) Velocity vs. time [25C2] (3) Housing/stock price vs. time [25C3] (4) Temperature vs. time [25C4]	

Regarding the items related to the quadratic functions /equations /inequalities (Items 19, 21, 22, & 23), the analysis was focused on the representations used and the shift between different representations (which will be further illustrated in the Results section).

Moreover, the concept of flexibility was defined by adopting different perspectives and different representations. Each shift between representations is coded as an event of flexibility if the participant is successful in solving the problem through this shift (e.g., score 3 or 4). For example, in item 22, the participants' responses can be categorized as three types, and each type presents a flexible event (see table; 3.9).

Table 3.9. Categories and Flexibility with Regard to Item 22

Solutions /Mistakes	Flexibility
Solution	
The effects of changing of a, b and c on the changes of the graphs of quadratic function.	Yes (graph vs. algebra)
Symmetrical line $x = -\dfrac{b}{2a}$, a is not changed, then only b need to changed.	Yes (graph vs. algebra)
Based on the algebraic relationship g(x)=f(-x), finding the coefficients of g(x) (a1=a, b1=-b, and c1=c).	Yes (graph vs. algebra)
Mistakes	
Based on $g(x)= a(x - h)^2 + b(x - h) + c$, make a statement that at least two of three (a, b, and c) need to be changed	No
According to $x = -\dfrac{b}{2a}$, $y = \dfrac{4ac - b^2}{4a}$, make a statement that at least two of three (a, b, and c) need to be changed.	No

Another example, in item 23 the problem could be solved in two steps: finding the quadratic function and then finding the maximum. First, three forms of quadratic formula methods: $y = ax^2 + bx + c$ (FM1); $y = a(x - x_1)(x - x_2)$ (FM2); and $y = a(x - h)^2 + k$ (FM3) can be used to find a quadratic function expression. Then, the student could have used three possible methods to find the maximum value: (1) transforming into $y = a(x - h)^2 + k$, then finding the maximum (MM1); (2) using formula $x = -\dfrac{b}{2a}$, $y_{max\,imum} = \dfrac{4ac - b^2}{4a}$ (MM2); and (3) taking the derivative: set $\dfrac{dy}{dx} = 0$, then find x=1, $y_{max\,imum} = f(1)$ (MM3).

All methods of solving the question are the combinations of the above methods as follows:

1. FM1 (step 1) and MM1 (step 2);

2. FM1 (step 1) and MM2 (step 2);

3. FM2/FM3 (step 1) and MM1 (step 2);

4. M2/FM3 (step 1) and MM2 (step 2);

5. FM1/FM2/FM3 (step 1) and MM3 (step 2).

As far as the shifts between representations are concerned, I coded one event of demonstrating flexibility for each of the methods 1 to 4, but not for method 5. In the first four methods, it is necessary to shift from a different quadratic formula.

3.3.4 Quantitative Analysis

I analyzed the quantitative data from three aspects. First, I analyzed the item mean and performed a t-test detecting mean differences between China and the U.S. Then, I analyzed the relationships between different variables (including latent variables) by a path model analysis and the fitness of the theoretical model of KAT by estimating instrument models. Third, I analyzed the correlation between the flexibility and other variables.

3.3.5 Interview Data Analysis

The U.S. interview data was analyzed to further illustrate prospective teachers' responses (their thoughts) to open-ended items. It is aimed at providing more detailed interpretation of participants' answers.

3.4   Framework for Data Analysis

The quantitative data analysis results were further illustrated and interpreted by the qualitative findings. Figure 3.2 describes the entire process of data analysis. According to this diagram, the items were first quantified into quantitative data for item and construct analysis. With regard to item analysis, the item mean was analyzed and compared using SPSS 16.0, and the path analysis and instrument model estimation were conducted by AMOS 16. In addition, a correlation analysis was used to investigate the relationship between flexibility and other variables, such as different knowledge components.

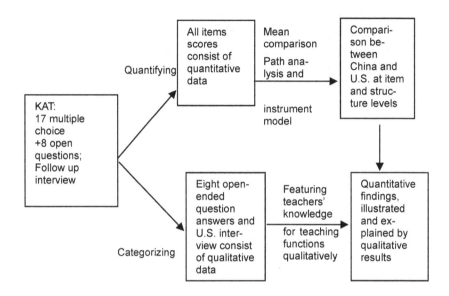

Figure 3.2. Process of data analysis.

With regard to the eight open-ended questions on teacher's knowledge of algebra for teaching, a qualitative analysis was performed. The purpose was to identify the characteristics of teachers' knowledge for teaching the concept of function cross-culturally. A particular focus was put on the strategies used and flexibility in adapting perspectives of function concept and selecting representations. The follow up interview data analysis was used for further clarifying participants' knowledge for teaching the concept of function. Finally, the qualitative results were further used for interpreting quantitative findings and the conclusions of the study were made.

3.5   Conclusion

In this chapter, I described the design of study, which is a mixed method by using a questionnaire. The process of development of the instrument for this study was described and discussed. Next, the data collection procedures in China and the U.S. were described. After that, the methods of data analysis were illustrated in detail. Finally, I summarized the strategies to integrate the findings based on quantitative and qualitative analyses to make conclusions of the study.

4 Chapter Four: Results

The findings of this study are organized into four sections. First, I report comparative results of KAT at item and structural levels between China and the U.S. Second, I present the relationship among background variables and components of KAT in China and the U.S. Third, I compare the similarities and differences of KTCF between China and the U.S. Fourth, I present an analysis of correlation between flexibility and other variables. Finally, I summarize the findings of the study with regard to the four research questions.

4.1 Comparison of KAT between China and the U.S.

4.1.1 Reliability of the Instrument

The questionnaire is designed to measure three types of knowledge: school mathematics, advanced mathematics and teaching mathematics. Each item belongs to one of the three categories. The distribution of items to different categories is shown in Table 4.1.

Thus, there are 7 items (1, 3, 6, 14, 17, 19, & 23) in *school mathematics*, 8 items (4, 8, 9, 12, 13, 16, 20, & 24) in *advanced mathematics*, and 10 items (2, 4, 7, 10, 11, 15, 18, 21, 22, & 25) in *teaching mathematics*. The reliabilities (Cronbach's Alpha) of the instrument are .88 (N=491, the whole sample), 0.61 (N=115, the U.S. sample), and .73 (N=376, the Chinese sample).

Table 4.1. Category of Knowledge of algebra for teaching

Items	Types of knowledge	Items	Types of knowledge
1	1	14	1
2	3	15	3
3	1	16	2
4	2	17	1
5	3	18	3
6	1	19	1
7	3	20	2
8	2	21	3
9	2	22	3
10	3	23	1
11	3	24	2
12	2	25	3
13	2		

Note. 1-School mathematics; 2-Advanced mathematics; 3- Teaching mathematics.

4.1.2 The Mean Differences of Items and Components between China and the U.S.

First, I compared the mean differences of multiple-choice items. The mean and t-test values are listed in Table 4.2.

Table 4.2. Mean Difference of Multiple Choices between China and the U.S.

	Mean		T-Test
Item	China	U.S.	
1	.90	.85	1.42
2	.96	.57	8.38**
3	.96	.89	2.43**
4	.78	.15	16.06**
5	.95	.18	20.70**
6	.38	.30	1.66

	Mean		T-Test
Item	China	U.S.	Item
7	.83	.57	5.20**
8	.90	.16	19.95**
9	.47	.20	5.90**
10	.68	.66	.45
11	.62	.78	-3.43**
12	.47	.27	4.01**
13	.64	.23	8.56**
14	.87	.40	9.64**
15	.66	.22	9.73**
16	.92	.29	14.19**
17	.96	.52	9.23**

Note. **$p<0.001$

This table showed that Chinese participants performed significantly higher than the U.S. counterparts, except for four items (1, 6, 10, and 11). On one item (11), the U.S. participants achieved a significantly higher mean score than Chinese counterparts (Mean difference=0.16, t=3.43, p<0.001). On items 1, 6, & 10, although the Chinese participants scored higher than their U.S. counterparts, there was no significant difference. On all remaining items, Chinese participants' mean scores were significantly higher than the U.S. counterparts (p<. 001). In addition to the above-mentioned items (1, 6, 10, and 11), five items (4, 5, 8, 15, and 16), with significant differences between China and the U.S. are further discussed in detail in the next section.

With regard to the open-ended items and components of KAT, the means and tests of significance are displayed in table 4.3.

Table 4.3. Mean Difference of Open-ended Items and Components of KAT between China
and the U.S.

	Mean		t-Test
Item	China	U.S.	
18	2.91	1.51	10.21**
19	3.66	.76	34.62**
20	3.47	.79	21.29**
21	2.97	.18	31.60**
22	2.64	1.40	8.44**
23	3.29	.29	33.01**
24	3.25	.02	46.63**
25	2.23	1.43	5.53**
SM	11.03	4.00	39.19**
AM	10.89	2.1	42.36**
TM	15.45	7.50	21.04**
KAT	37.37	13.60	39.40**

Note. SM-School mathematics; AM -Advanced mathematics; TM -Teaching mathematics;
and KAT - Knowledge of algebra for teaching.

On all the open-ended items, there are significant mean differences between China and the U.S. Chinese participants achieved significantly
higher means in school mathematics (mean difference=7.03, t=39.19,
p<.001), advanced mathematics (MD=8.78, t=42.36, p<.001), and teaching mathematics (MD=7.95, t=21.04, p<.001) than U.S. counterparts.
Consequently, Chinese participants achieved significantly higher mean
of KAT (MD=24.77, t=39.40, p<.001) than their U.S. counterparts.

Since the Chinese sample is relatively large, I did a multiple comparison of KAT with regard to different institutions. All of the participating
universities were classified into three types based on a 2009 university
league list in China. One high achieving university (type 1), two interme-
diate achieving universities (type 2), and two low achieving universities

(type 3) (See Table 3.2 in Chapter 3 for details). It was found that there was no significant mean difference between type 1 and type 2 universities, but type 1 and type 2 universities had a significantly higher mean score of KAT than type 3 universities. Moreover, given the fact that only junior or senior in-service teachers came from type 2 universities, I excluded participants from type 2 universities for qualitative analysis. Thus, in this study, the high-achieving group consists of participants from type 1 (N=109) and the low-achieving group consists of participants from type 3 universities (N=147).

4.1.3 Analysis of Selected Multiple Choice Items

In this part, I examine several multiple-choice items in detail. These items include: item 11, in which U.S. participants outperformed Chinese counterparts; three items, 1, 6, & 10, in which there are no significant mean differences between China and the U.S.; and five items, 4, 5, 8, 15 & 16, in which the means of participants from China were significantly higher than those from the U.S. I made a comparison between the China high-achieving group (N=109), the China low-achieving group (N=147), and the U.S. (N=115) in order to better understand how the participants answered these questions.

Item 11, in a first year algebra class, which of the following is **NOT** an appropriate way to introduce the concept of slope of a line?

A. Talk about the rate of change of a graph of a line on an interval.

B. Talk about speed as distance divided by time.

C. Toss a ball in the air and use a motion detector to graph its trajectory.

D. Apply the formula $slope = \dfrac{rise}{run}$ to several points in the plane.

E. Discuss the meaning of *m* in the graphs of several equations of the form

$y = mx + b$.

The distribution of different choices of this item is displayed for China (high achieving vs. low achieving groups) and the U.S. sample in Table 4.4. (C is the correct choice).

Table 4.4. The Choice Distribution of Item 11 in China and the U.S

| | China | | U.S. |
	High (%)	Low (%)	(%)
A	11	8	8
B	11	11	5
C	64	61	78
D	12	20	5
E	2	1	6

More than 60% of Chinese participants identified the correct choice and the Chinese participants also made a variety of wrong choices. These results may imply the Chinese participants are not familiar with using various learning situations (particular contextual situations) to introduce the concept of slope. It may reflect that the Chinese participants' learning experience was limited in mathematical context. It may also reflect that they may have memorized the formula of slope but do not understand the geometrical meaning of the formula (12% from high achieving group and 20% for low-achieving group were not able to use the formula of *slope=rise/run* in China). On the other hand, the U.S. participants had a high rate of correct choice (78%). This may imply that the U.S. participants had a better understanding of the concept and were exposed to multiple contextual situations.

Item 1, at a storewide sale, shirts cost $8 each and pants cost $12 each. If S is the number of shirts and P is the number of pants bought, which of the following describes the expression 8S + 12P?

A. The number of shirts and pants bought

B. The cost of 8 shirts and 12 pants

C. The cost of P shirts and S pants

D. The cost of S shirts and P pants

The different choices of the item 1 are displayed in Table 4.5 (Correct choice is C)

Table 4.5. The Choice Distribution of Item 1 in China and the U.S

	China		U.S.
	High (%)	Low (%)	(%)
A	0	2	1
B	4	6	9
C	3	3	4
D	92	89	86

The Table showed both Chinese and U.S. participants made a high rate of correct choice D (greater than 86%). Only a small part of them made a wrong choice of B or C. The result may imply both Chinese and U.S. participants are familiar with presenting quantitative relationship by using algebra expressions.

Item 6, which of the following can be represented by areas of rectangles?

i. The equivalence of fractions and percents, e.g. $\frac{3}{5} = 60\,\%$

ii. The distributive property of multiplication over addition:

For all real numbers a, b, and c, we have $a(b + c) = ab + ac$

iii. The   expansion   of   the   square   of   a   binomial:
$(a+b)^2 = a^2 + 2ab + b^2$

A. ii only

B. i and ii only

C. i and iii only

D. ii and iii only

E. i, ii, and iii

The different choices of the item 6 are displayed in Table 4.6 (Correct choice is E)

Table 4.6. The Choice Distribution of Item 6 in China and the U.S.

	China		US
	High (%)	Low (%)	(%)
A	3	3	7
B	3	15	16
C	4	14	26
D	56	35	22
E	35	34	30

This table showed that both Chinese and U.S. participants scored very low (between 30% to 35%), and had no significant mean difference between China and the U.S. Interestingly, more than half (56%) of the Chinese participants from high-achieving group made choice D. That means they did not realize that the equation $\frac{3}{5} = 60\,\%$ could be represented by the area of a rectangle (Choice D). More than one third of participants from low achieving group and about one fourth of U.S. participants made the same choice (Choice D). In addition, about one-fourth U.S. participants did not realize that $a(b + c) = ab + ac$ can be represented by the area of a rectangle (Choice C). In summary, both Chinese and U.S. participants were weak in using geometrical representation to present fraction/percentage and algebraic formula. That means the participants from both countries were not skilled in linkage between algebraic (arithmetic) and geometrical representations.

Item 10.   A textbook includes the following theorem:

If line l_1 has slope m_1 and line l_2 has slope m_2 then $l_1 \perp$

l_2 if and only if

$m_1 \cdot m_2 = -1$ *(i.e. "slopes are negative reciprocals").*

(McDougal Littell, Algebra 2)

Three teachers were discussing whether or not this statement generalizes to all lines in the Cartesian plane.

Mrs. Allen: The statement of the theorem is incomplete: it doesn't provide for the pair of lines where one is horizontal and one is vertical. Such lines are perpendicular.

Mr. Brown: The statement is fine: a horizontal line has slope 0 and a vertical line has slope ∞ and it's okay to think of 0 times ∞ as -1.

Ms. Corelli: The statement is fine; horizontal and vertical lines are not perpendicular.

Whose comments are correct?

    A. Mrs. Allen only

    B. Mr. Brown only

C. Ms. Corelli only

D. Mr. Brown and Ms. Corelli

E. None are correct.

The different choices of the item 6 are displayed in Table 4.7 (Correct choice is A)

Table 4.7. The Choice Distribution of Item 10 in China and the U.S.

	China		U.S.
	High (%)	Low (%)	(%)
A	70	63	66
B	7	4	10
C	1	3	4
D	1	3	4
E	21	28	16

Both Chinese and U.S. participants had a similar correct rate (about 67%). It is interesting that about one quarter of participants in the two countries did not agree with that given explanation (Choice E). Also, there was a small number of participants that agreed "0 times -∞ as −1" by choosing B. This result alerts that it is in need to introduce a theorem more rigorously.

Since items 4, 5, and 16 are related to irrational function, irrational equation and derivative of polynomial, Chinese participants outper-formed U.S. counterparts significantly. That means Chinese participants achieved high-scoring in advanced algebra computation. I further exam-ined two other items, 8 and 15, on which Chinese participants performed very well.

Item 8.   The given graph represents speed vs. time for two cars. (Assume the cars start from the same position and are traveling in the same direction.) Use this information and the graph below to answer.

What is the relationship between the *position* of car A and car B at t = 1 hour?

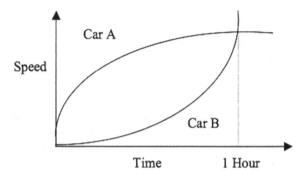

Figure 4.1. Diagram of the relationship between speed and time.

A. The cars are at the same position

B. Car A is ahead of car B

C. Car B is passing car A

D. Car A and car B are colliding

E. The cars are at the same position and car B is passing car A.

The different choices of the item 8 are displayed in Table 4.8 (Correct one is B)

Table 4.8 The Choice Distribution of Item 8 in China and the U.S.

	China		U.S.
	High (%)	Low (%)	(%)
A	0	2	19
B	96	88	16
C	1	2	13
D	1	2	4
E	2	5	48

The Chinese participants did extremely well in this item. The average correct rate is 90%. Even participants from the low-achieving group had about 88% correct rate. In order to get a correct answer, it is required to have a logical reasoning based on speed and time relationship and the graph. Only a small part of the participants from the low-achieving group made their judgments based on the visual information without having an appropriate understanding of the meaning of speed vs. time graph (Choice A & E).

In the U.S. participants, only 16 % of them made a correct choice based on logical reasoning and graphical representation. About half of them (48%) made a wrong choice based on visual information only: intersection point and high over (Choice E) or partially using the visual information: intersection point (Choice A) or high over (Choice B). It may be that many U.S. participants used visual judgment rather than logical reasoning.

Item 15. Which of the following (taken by itself) would give **substantial** help to a student who wants to expand $(x + y + z)^2$?

i. See what happens in an example, such as $(3 + 4 + 5)^2$.

ii. Use $(x + y + z)^2 = ((x + y) + z)^2$ and the expansion of $(a + b)^2$.

iii. Use the geometric model shown below.

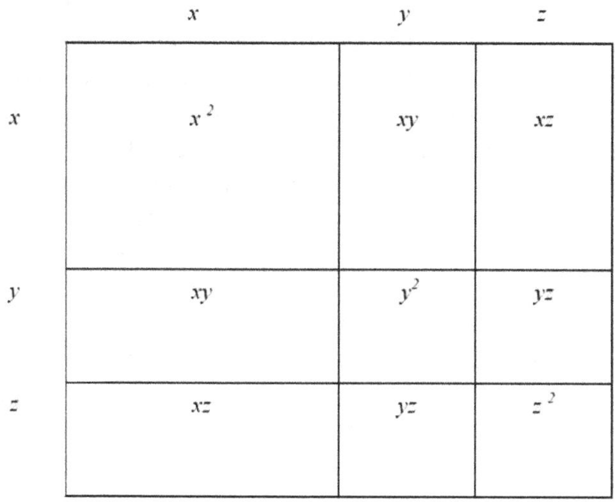

	x	y	z
x	x^2	xy	xz
y	xy	y^2	yz
z	xz	yz	z^2

Figure 4.2. Diagram of expansion of $(x+y+z)^2$

A. ii only

B. iii only   **C.** i and ii only

D. ii and iii only

E. i, ii and iii

The different choices of the item 15 are displayed in Table 4.9 (Correct choice is D)

Table 4.9. The Distribution of Choice of Item 15 in China and the U.S

	China		U.S.
	High (%)	Low (%)	(%)
A	6	11	2
B	6	5	57
C	3	3	1
D	73	72	22
E	13	19	15

The table showed that about 72% of the Chinese participants made the correct choice. They realized that using algebraic computation and geometrical model can help students to understand the algebraic expansion. About 15% of the Chinese participants believed the numerical computation can also be helpful. However, more than half of U.S. participants believed only the geometry mode is helpful, while only 35% of the participants recognized the usefulness of exploring algebraic expression. About 16% of U.S. participants believed the usefulness of exploring numerical computation. Thus, the U.S. participants relied on the geometrical model to reason while the Chinese counterparts make their reason based on the geometrical model and algebraic computation.

In summary, the analysis of these purposely selected items show that both Chinese and U.S. participants scored high in expressing contextual situations using algebraic expressions (item 1) , and revealed a weakness in linking multiple representations such as numerical, algebraic and geometrical ones (item 6). Compared with the Chinese participants, the U.S. counterparts demonstrated strengths in understanding the concept of slope from different aspects (item 11). However, when making reasoning or judgment based on a graph, the U.S. participants often preferred to rely on visual or geometrical information, while the Chinese counterparts tended to make logical reasoning with the support of algebraic and geometrical representation and computation (items 8 & 15).

4.2   Relationship among Components of KAT in China and the U.S.

In this section, I report path models in China and the U.S., and a measurement model in China. The relationships among background variables and different components of KAT are analyzed based on these models.

4.2.1 Path Model Analysis

In this part, I examine the relationship between background variables (course taking and grade) and components of KAT, and relationships between components of KAT. Researchers (e.g., Monk, 1994) suggest the number of courses taken by teachers is positively related to how much their students learn in mathematics at the secondary level. Thus, in this study, I created a conceptual frame for examining the relationship between background variables, including the number of mathematics and mathematics education courses taken, and the components of KAT as shown in Fig. 4.3.

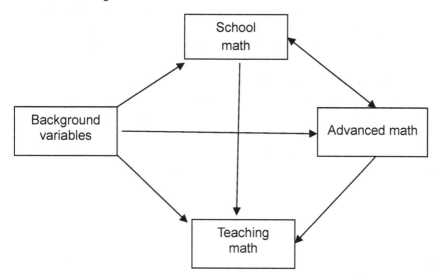

Fig. 4.3. Theoretical model relating to background variables and components of KAT.

As outlined in Fig. 4.3., the relationship between background variables (the number of math courses taken in high school, the number of courses taken in college, and the grade), and components of KAT (SM, AM and TM) were modeled using a series of path models (See Fig 4.4). It was assumed that SM and AM interact with each other and that both of them impact TM. Furthermore, the KAT (including SM, AM and TM) is hypothesized to be a function of background variables. SM, AM and TM are considered as endogenous variables in the model. The aim of the path analysis is to include the entire variables that may contribute to the explanation of the variance in the endogenous variables. As it is impossible to include everything that may impact these variables, an error term is included in the model to be estimated for each endogenous variable (e.g., res, rea, and ret are the error terms for SM, AM, and TM respectively, see Fig 4.4). The errors reflect all those unobserved predictors that were not measured in this study nor included in this model.

In addition, this system of variables was hypothesized to be influenced by participants' background characteristics; including the number of high school math courses taken, the number of college math education courses taken and grade level. These variables are considered to be exogenous (i.e., independent variables that are not dependent on or predicted by any other variables in the model). The variables are co-varied to influence KAT.

I estimated this mode by using AMOS 16. Initially, all the parameters for the background variables on the system of variables were estimated. Consequently, to achieve a good fit, some paths were deleted that were not significant. The final model is presented in Fig 4.4. In this diagram, the bold arrow lines represent a significant effect while the dashed lines represent a non-significant effect. Estimates are in raw score form.

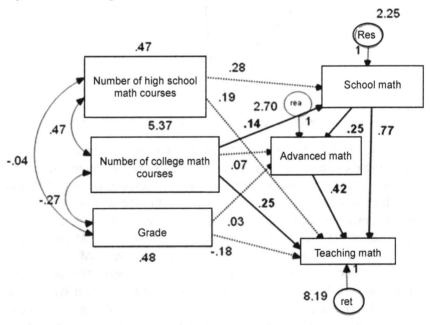

Fig. 4.4 . Final path model of the course taking, grade level, and KAT in the U.S.

In this model, the chi-square test for lack of fit was not significant, $\chi^2(2) = 0.08$, p=0.96. This means the data has a good fit (Hu & Bentler, 1999). Moreover, other fit indices showed there was a good fit. The comparative fit index CFI equals 1. This index can take on a value from 0 to 1 with values closer to 1 showing a better fit and value greater than .90 usually indicating a relatively good fit (Kline, 2005). The root mean square error of approximation (RMSEA) was 0.00. This index takes into account the complexity of the model, and it can range from 0 to 1 with less than 0.05 of presenting a good fit.

The parameters shown in Fig. 4.4, school mathematics were found to have direct and significant effects on teaching mathematics (β=0.77, p<0.05) and advanced math (β=0.25, p<0.05). Number of college math courses was found to have a significant effect on school math (β=0.14, p<0.05) and teaching math (β=0.25, p<0.05) while advanced math was found to have a direct and significant effect on teaching math (β=0.42, p<0.05). However, the number of high school math courses and grade level were not found to have significant effects on school math (β=0.28) and teaching math (β=0.19), and the number of college math courses was not found to have significant effect on advanced math (β=0.07).

Similarly, a final path model of the course taking, grade level, and KAT in China was created as Fig. 4.5. The dashed lines represent non-significant effects while the bold lines represent significant effects. Since there was no number of courses taken in high school in China, we just focused on the number of courses taken in college.

In this model, the chi-square test of lack of fit was not significant, $\chi^2(1) = 0.583$, p=0.45. This means the data has a good fit. Moreover, other fit indices shown the model fit is good (CFI=1, RMSEA=0.000). The parameter estimates are shown for the Chinese model (see Fig. 4.5), the number of college math education courses was found to have a direct and significant positive effect on advanced math (β=0.10, p<0.05) while the grade level was found to have a direct and significant negative effect on school mathematics and advanced math (β=-0.92, p<0.05).

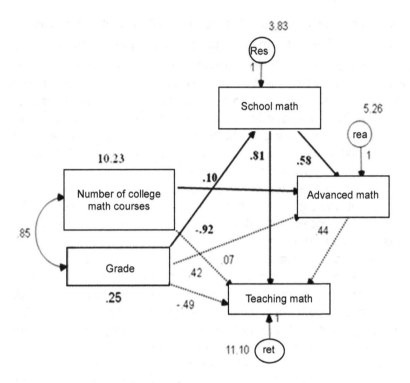

Fig. 4.5 . Final path model of the course taking, grade level, and KAT in the China.

School mathematics was found to have significant positive effects on advanced math (β=0.58, p<0.05) and teaching math (β=0.81, p<0.05). However, grade levels were not found to have an effect on advanced math (β=0.42) or teaching math (β=-.49). Number of college math courses was not found to have an effect on school math (β=0.07), nor was advanced math found to have an effect on teaching math (β=0.44).

Comparing the Chinese model and the U.S. model, the number of college math courses was not found to have an effect on advanced math (β=0.07) in the U.S., while in the effect was significant (β=0.10, p<0.05) in China. In addition, in the U.S. sample, number of college math courses taken was found to have a significant effect on teaching mathematics (β=0.25, p<0.05), while in the Chinese sample, the effect was not significant (β=0.07). This result may indicate that the Chinese teacher preparation emphasizes content knowledge while the U.S. teacher preparation emphasizes pedagogical knowledge. When considering the size of effect between different components in China and the U.S., it was found that there were stronger correlations in China than those in the U.S. (For example, the effect of school mathematics on teaching math is 0.81 in China, while the corresponding effect is 0.77 in the U.S.; the effect of school mathematics on advanced math is 0.58 in China, while the corresponding effect is 0.25 in the U.S.). However, the differences of correlations (See Table 4.10) are not significant between China and the U.S. based on the Fisher's Z test.

Table 4.10. The Correlations of Different Components of KAT

	U.S. (N=115)			China (N=376)		
	SM	AM	TM	SM	AM	TM
SM	1	.242**	.449**	1	0.423**	.522**
AM	.242**	1	.336**	.423**	1	.449**
TM	.449**	.336**	1	.522**	.449**	1

In this model, the chi-square test of lack of fit was significant, $\chi^2(174) = 245.34$, p<0.001. This showed the data (p<0.05) did not have a good fit to the model. However, other fit indices shown a relatively good fit. The Comparative Fit Indices (CFI) equaled 0.914 (>0.90) and the Root Mean Square Error of Approximation (RMSEA) was 0.03 (0.02-0.04) (<0.05). These two indexes indicate the Chinese data fit the model relatively well.

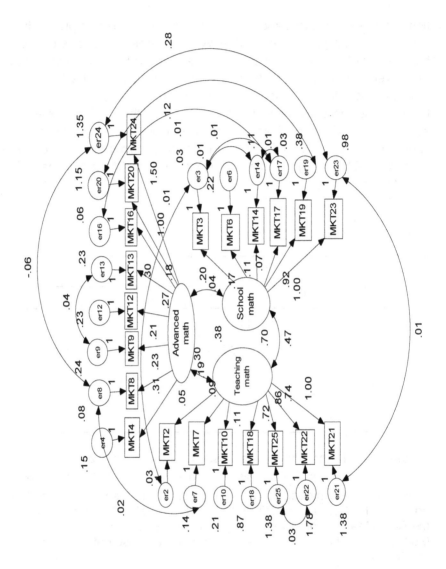

Fig 4.6. The final model of KAT in China.

The bidirectional relationships between different observed variables are modeled by allowing the error terms for each of these variables to covary (for example, labeled er 21 and er 23). Corvariance of error terms essentially reflects the correlation between two variables.

Estimates of model parameters were obtained using AMOS 16 (Byrne, 2010). Initially, all parameters of the theoretical model were estimated. Consequently, based on the inspection of weight load (larger than .02) and model fit indexes, the model was re-estimated. The final model with all parameter estimates is presented in Fig. 4.6.

The standardized regression weights (larger than 0.10) of all the observed variables in the final model were displayed in Table 4.11.

Table 4.11. Standardized Regression Weights Estimate

Standardized regression weight			Estimate
MKT3	<---	SM	.147
MKT6	<---	SM	.215
MKT14	<---	SM	.211
MKT19	<---	SM	.677
MKT23	<---	SM	.530
MKT17	<---	SM	.243
MKT10	<---	TM	.206
MKT7	<---	TM	.196
MKT25	<---	TM	.522
MKT18	<---	TM	.544
MKT21	<---	TM	.579
MKT22	<---	TM	.420
MKT2	<---	TM	.211
MKT9	<---	AM	.185
MKT12	<---	AM	.237
MKT13	<---	AM	.163
MKT16	<---	AM	.474
MKT20	<---	AM	.374
MKT4	<---	AM	.323
MKT24	<---	AM	.488

This table showed that except for four items (MKT13, MKT9, MKT7, MKT3), the others had weights greater than 0.20. Among these items, the open-ended items had relatively greater weights ranging from 0.420 to 0.677.

In addition, the correlations between latent variables (components of KAT) are shown in the following table 4.12. The correlations of these latent variables were relatively high (.75 to .91).

Table 4.12. The Correlation and Co-variance of Different Variables

Correlations	Estimates
SM <--> TM	.91
SM <--> AM	.75
TM <--> AM	.83

This table shows that there was a high correlation between school mathematics and teaching mathematics (r=0.91). Compared with the correlation between school mathematics and advanced mathematics (r=0.74), the correlation between teaching mathematics and advanced mathematics (r=0.83) was relatively high.

The final Chinese model of KAT showed that there are many links between errors, within the same component or across components. These links imply that these observed variables are not able to be used to measure different components of KAT exclusively, rather they are used to jointly measure an interconnected structure of knowledge of algebra for teaching. When examining the linked items themselves, some of them are essentially related. For example, item 7 is about linear function and its graph while Item 8 is about using a graph of speed vs. time to interpret a daily life situation. For another example, item 3 is about polynomial function computation while item 14 is about irrational equation solution. However, in other cases, the items are not obviously related. For example, item 19 is about solving quadratic inequality while item 20 is about matrix computation. Item 23 is about finding quadratic function and its maximum while item 24 is about proving a proposition of sum of two linear functions. The model seems to suggest that the theoretically and artificially exclusive components of KAT are essentially interconnected. That means that knowledge for teaching should be treated as a comprehensive and interconnected entity and construct.

However, with the U.S. data, the model is not admissible. It may be due to the small sample size (N=115 does not meet the minimum requirement of 10 × 25=250 cases required for a full estimation).

4.3   Comparisons of KTCF between China and the U.S.

In this part, I presented a detailed analysis on the participants' responses to the open-ended items in terms of their overall performance, typical strategies/methods, and misconceptions/mistakes. With the U.S. case, I complemented relevant analysis with the interview data. Rather than analyzing items one by one in order, I grouped these items into three categories: One item is related to matrix and logical inference (Item 20); Three items are related to function concept in general (Items 18, 24,& 25), and the remaining four open-ended items are related to quadratic functions/ equations/ inequalities (Items 19, 21, 22, & 23) in particular.

For the matrix item, the analysis was focused on the logical equivalence, and matrix computations. For the items related to the concept of function, the analysis was focused on the perspectives of the function concept (process vs. object); for the items related to quadratic functions/equations/inequalities, a two dimension framework consists of the perspective of function, and the flexibility in the use of representations guides the data analysis. The framework is aimed to capture the understanding of the function concept and the flexibility in using representation (using verbal, tabular, algebraic, and graphical representations, translation between different representations and transformation within a representation) (see Table 3.8).

In the sessions that follow, I will present relevant findings in these three areas.

4.3.1 Logical Reasoning in Matrix System

Item 20 is an open-ended item used for measuring advanced knowledge. As Table 4.3 showed that there was a significant mean difference between China and the U.S. (Mean difference=2.68, t=21.29, p<0.001). Moreover, the score distribution is displayed in Fig. 4.7.

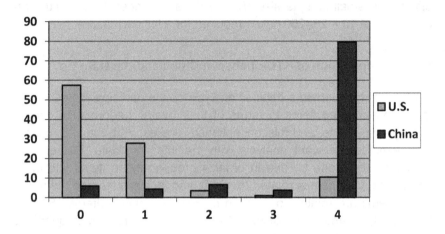

Fig. 4.7. Score distribution of item 20.

This figure shows that more than 80% of Chinese participants provided correct proofs while only about 10% of their U.S. counterparts did the same. Less than 6% of Chinese participants did not provide any useful information while more than half (57.4%) of the U.S. participants did the same.

The item 20 is as follows:

Let A=$\begin{bmatrix} p & q \\ r & s \end{bmatrix}$ and $B = \begin{bmatrix} t & u \\ v & w \end{bmatrix}$, Then $A \Delta B$ is defined to be $\begin{bmatrix} pt & qu \\ rv & sw \end{bmatrix}$. Is it true that if $A \Delta B = O$, then either A = O or B = O

(where O represents the zero matrix)? Justify your answer and show your work in the Answer Booklet.

In fact, the participants are required to provide a counterexample to disprove the statement. However, a common misconception of the U.S. participants was to use the wrong logical reasoning as follows: $p \to q \Leftrightarrow q \to p$, namely, using the following logic "if A=0, then AΔB=0 or if B=0, then AΔB=0" to prove: "if AΔB=0, then A=0, or B=0". More than a quarter of the U.S. participants made this mistake, while only a few of the Chinese counterparts made the same mistakes.

A few of the U.S. participants inappropriately generalized the same proposition from real number systems, namely $x.y = 0; x = 0$, $y = 0$. (x, y are real numbers) to matrix system. No Chinese participants made this overgeneralization.

U.S. participants' interpretation. The interviews with U.S. participants further confirmed their difficulties in providing a correct proof. Two of the five interviewees (Jenny and Stacy) gave the correct answer with an appropriate counterexample. For example, Stacy explained why she tried to disprove the statement as follows: "if someone wants to prove a proposition, s/he has to provide the whole process of proving. However, if someone just wants to disprove a proposition, s/he only provides a counterexample, so, I considered to find a counterexample". The others gave a wrong judgment by providing examples such as "A=0, then $A\Delta B = O$" or "B=0, then $A\Delta B = O$" by "guess and check". However, when the researcher asked them "whether it is possible that if A $\neq$ 0, B $\neq$ 0, but $A\Delta B = O$", two of them (Larry and Alisa) took a second thought, and found a counterexample to disprove the statement. For example, Alisa gave a counterexample, A= $\begin{bmatrix} 0 & 2 \\ 1 & 0 \end{bmatrix}, B = \begin{bmatrix} 3 & 0 \\ 0 & 4 \end{bmatrix}$. However, Kerri was still struggling with finding a counterexample by saying "it is a trick".

In summary, more than three-fourths (85%) of the U.S. participants were not able to provide any relevant information, and about one fourth were confused with the logical proposition relationship between "$p \to q$" and "$q \to p$".

4.3.2 Flexibility in Adopting Perspectives of Function Concept

The items 18, 24 and 25 are particularly used for measuring the know-ledge of understanding and applying function concept from different per-spectives (*process* and *object*). It is crucial for participants to provide correct answers and explanations if they select the appropriate perspec-tive. Item 18 is in favor of using process perspective; item 24 is easily proved if adopting an object perspective. It is necessary to connect those two perspectives when solving item 25.

Responses to Item 18. There was a significant mean difference of item 18 between China and the U.S. (mean difference=1.4, *t*=10.21, *p*<0.001, see Table 4.3). The score distribution of the item is shown in Fig. 4.8.

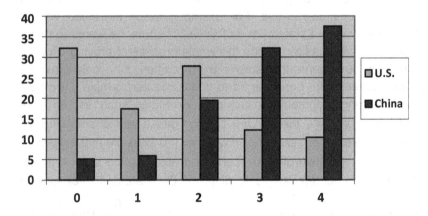

Fig. 4.8. Score distribution of Item 18.

The table showed that about 23% of U.S. participants got a correct an-swer (10.4%) or almost correct answer with minor mistakes (12.2%), while there are about 70% of the Chinese counterparts did the same. The following are samples of correct answers in China and the U.S. (Fig. 4. 9)

Fig 4.9. Examples of answers to item 18 in China and the U.S.

The Chinese participant directly used the definition of function: Let two real number sets A, B, if for any a value *x* in set A, there is only one value *b* in set B corresponding to *a*, then this corresponding relationship *f* from set A to set B is a function. According to this definition, (i) and (ii) are functions. However, the U.S. prospective teacher used a diagram to visualize the function relationship and then made a judgment of these two given relations

Conversely, 32% of the U.S. participants' provided nothing or meaningless information about the answers of the item 18; only about 5% Chinese did the same. About 28% of U.S. participants and 19% of Chinese participants just gave correct answers without any interpretations or give one correct answer and relevant explanations.

Perspectives adopted in Item 18. In addition, the perspectives used in participants' interpretation are listed in Table 4.13.

Table 4.13. Perspectives Adopted in the Responses to Item 18

Perspective	Description	Frequency		
		China		U.S.
		High (%)	Low (%)	(%)
Process	Corresponding relationship between domain and range (one-to-one/multiple-to-one)	51	31	6

Object	Algebraic expressions (constant value; two expressions) Graphic features (one line, many holes/un-continuous)	12	10	9

With regard to this item, it is more appropriate to adopt a process perspective. In China, in the high achieving group, more than half (51%) adopted the process perspective and in the low achieving group, more than one-third of the participants (31%) adopted this perspective. However, the U.S. participants preferred using object perspective (9%), namely, basing on function expressions and graphic features to using essentially corresponding relationship features (6%).

U.S. participants' interpretation in item 18. In responding to how they made their judgments, they reported using the vertical line test (Larry, Alisa, and Stacy) or diagrams presenting corresponding relationship between two sets (Kerri). Jenny made her wrong judgment based on visual graphical images. Since she had a difficulty in drawing the graph of the second relation, she believed it was not a function. However, when asked whether she heard of the vertical line test, she clearly stated that "one x value can only have one corresponding y value; one x value cannot be corresponded to two y-values." Kerri said she "is a visual learner, and likes using diagrams representing the relationship between two sets (one-to-one or multiple-to-one, but not one-to-multiple)". Larry not only explained the vertical line test rule, but also showed an example ($x=y^2$) which cannot pass the vertical line test. Alisa and Stacy explained the rule by emphasizing, "each input [value] should have only one [corresponding] value, but that does not mean that different [input] values cannot have a same [corresponding] value."

With regard to students' mistakes, they attributed them to students' superficial understanding of the vertical line test (missing multiple x values correspond one y-value) or the confusion with "many holes", or the repeating output.

Four of them showed an accurate understanding of how to judge whether a relationship is a function or not based on corresponding relationships using either the vertical line test or diagrams to present the features of function relationship: one to one or multiple to one. They also realized that the "unusual" graphs of the function such as including constant value, many holes may confuse students' judgment.

Summary of responses to item 18. More Chinese than U.S. participants adopted process perspective, which protected them from the distraction of the unusual "appearances" of the function expressions or function graphs. These participants, who adopted a process perspective, were most likely to make a correct judgment, and provide reasonable explanations of those mistakes.

Responses to item 24. There was a significant mean difference of item 24 between China and the U.S. (MD=3.23, *t*=46.63, *p*<0.001). The score distribution of the item is shown in Fig. 4.8.

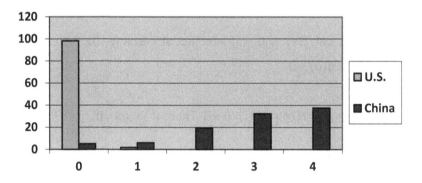

Fig.4.10. Score distribution of item 24.

There were two proofs as follows:

Method 1: Let $f(x)$ and $g(x)$ intersect at x-axis $(p, 0)$, then, the following statements are true:

(1) $f(p) = 0 \rightarrow ap + b = 0 \rightarrow p = -\dfrac{b}{a}$;

(2) $g(p) = 0 \rightarrow cp + d = 0 \rightarrow p = -\dfrac{d}{c}$;

(3) $f(p) = g(p) \rightarrow \dfrac{b}{a} = \dfrac{d}{c} \rightarrow ad = bc$;

(4) $f(p) = g(p) \rightarrow ap + b = cp + d \rightarrow p = -\dfrac{b+d}{a+c}$

According to the equation $(f + g)(p) = f(p) + g(p)$, and above statements, the participants could deduce $(f + g)(p) = 0$. Thus, $(f + g)(x)$ passes at point (p, 0).

Method 2. Let $f(x)$ and $g(x)$ intersect at x-axis (p, 0), then, $f(p) = 0$, $g(p) = 0$

So, $(f + g)(p) = f(p) + g(p) = 0 + 0$ Thus, $(f+g)(p) = 0$.

In method 1, the underlying method is to find the coordinate of the intersection point p and check whether $(f+g)(p) = 0$, while the strategy in method 2 is based on the definition of equation root and the definition of the sum of functions. Thus, the method 1 is mainly guided by the process perspective, while the method 2 is essentially guided by the object perspective.

The above figure showed that almost all U.S. participants gave up the effort to find a proof or provided some irrelevant statement. Only two of them gave some statements which were useful for developing a proof. On the other hand, in China, more than one-third of the participants provided a correct proof and the other one-third provided a rough correct proof with minor mistakes. About 5% of the Chinese participants left it blank, and another 6% just gave some related statements but failed to create a proof.

Perspective adopted in item 24. When looking at the perspectives or the strategies used in attempting to find a proof, the distribution of using different perspectives is shown in Table 4.14. The table showed that more than half of the participants (80% in high achieving group and 60% in low-achieving group) in China adopted the object perspective so that they can perform with functions as an entity and provide a proof effectively.

Table 4. 14. Perspectives Used in Item 24

Perspective	Description	Frequency		
		China		US(%)
		High (%)	Low (%)	
Process	Method 1	11	8	2
Object	Method 2	80	62	0

U.S. participants' explanations to item 24. Larry gave two concrete examples to explore the intersection points. However, in the interview, he used a general form of linear function, $f(x) = a_1 x + b_1$ and $g(x) = a_2 x + b_2$, and got a correct proof. Jenny just gave two concrete examples to explore, and then got stuck. The other three gave up their efforts to explain because they do not like proving.

Summary of responses to item 24. More than 60% of Chinese participants could prove roughly correct proofs (half of them with some minor mistakes). More importantly, they could adopt an appropriate perspective of function, namely, object perspective. Thus, they can operate on a function as an object so that they avoid the difficulty in finding the function expression itself. However, the U.S. participants simply gave up their attempts to find a proof.

Responses to item 25. There was a significant mean difference of item 25 between China and the U.S. (MD=0.80, t=5.53, p<0.001; See Table 4.3). The score distribution of the item is shown in Fig. 4.11.

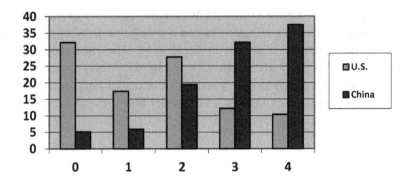

Fig. 4.11. Score distribution of item 25.

This figure showed that 45% of Chinese participants and about one-fifth (18%) of U.S. counterparts gave roughly correct answer and interpretation (3 or 4 scores). About one-third (28.4%) of Chinese participants and more than one-third (33%) of U.S. counterparts either gave correct explanations or gave an appropriate interpretation. One-fifth (19%) of Chinese participants and one-third (31%) of U.S. counterparts gave useless information. For example, the Chinese participants gave correct answers as follows:

> "The student's explanation is correct. He/she links the life situation with the graph, if the x-axis represents time and the y-axis represents the vertical height. The graph can also be explained as a car driving. At the beginning, the driver speeds up, then drives it at a constant velocity, after that slows down, and finally drives at a constant velocity."

> "If the graph is seen as a height above sea level (h) and the time (t), then the students' opinion is right. They just understand function at a visualization level;

> This graph can also be used to present the changes of the stock market. In the morning, the price of the stock is increasing, and keeps the same during the recession at noon. In the afternoon, the price of the stock goes down, and finally stops at the price the same as the price at the beginning of the market."

16 U.S. participants (14%) pointed out that the student's interpretation could be improved by denoting x-axis as time while the y-axis as height above sea level. For example, one participant described it as follows:

"That is a very creative answer, but [he/she] was not looking at the graph as a physical representation; we need to utilize it as a representation of data. The x-axis represents time while the y- axis represents height."

Thirty one U.S. participants (31%) gave the situation of speed over time to illustrate the same diagram. For example, one participant gave

"The graph could be showing speed vs. time where somebody is accelerating at an exponential rate, then goes a steady for period of time, and then slows at a constant rate, then stops."

Perspective adopted in item 25. It is necessary to have a connection of these two perspectives of function in order to find roughly correct answers. The different ways of explaining or interpreting the graph are displayed in Table 4.15.

Table 4.15. Different Ways of Interpretation the Graph

Types of interpretation	Frequency		
	China		U.S.
	High (%)	Low (%)	(%)
Height vs. Time	20	12	4
Velocity vs. Time	51	39	31
Housing/Stock Price vs. Time	2	0	0
Temperature vs. time	2	3	0
Distance vs. Time	6	5	4

The table shows that the majority of the participants explained, or interpreted the graph as a graph of a relationship between velocity and time. It seemed that the participants were in favor of a graph of velocity and time (51% in high achieving group in China, 39% in low achieving group in China and 31% in U.S. participants). The second frequent interpretation was the relationship between height and time (20% in high achieving group in China, 12% in low achieving group in China, and 4% in U.S. participants).

U.S. participants' interpretation to item 25. Three of the five interviewees (Larry, Alisa, and Stacy) realized that the original interpretation could be improved by pointing out the x-axis represents time while y-axis represents position (or height). All of them gave other examples of describing the diagram as the graph of speed over time, and two of them (Larry and Stacy) also mentioned about the graph of temperature over time. Two of them also mentioned they learned the graph of distance over time in one course called math and technology using CBR. (The CBR 2 is TI's answer to an easy, affordable data collection device! Designed for teachers who want their students to collect and analyze real-world motion data, such as distance, velocity and acceleration).

Summary of responses to item 25. The above analysis shows that compared with the U.S. participants, the Chinese participants were more likely to give correct interpretations, as well as give more diverse interpretations. For those who gave correct interpretations, it is necessary to have flexibility in shifting perspectives between process and object. Interview information further confirmed that U.S. participants generally demonstrated the appropriate knowledge about how to interpret the graph by using certain daily-life situations.

Summary of flexibility in selecting perspectives. The analysis of participants' responses to the three items suggests that compared with the U.S. participants, the Chinese participants demonstrated a flexibility in selecting appropriate perspectives of the function concept, namely, *process and object.* Moreover, Chinese participants provided more diverse interpretations than the U.S. interpretations.

4.3.3 Flexibility in Using and Shifting Different Representations

The items 19, 21, 22, and 23 are deliberately designed and used for measuring knowledge for understanding and applying quadratic functions/equations/inequalities through flexible use of multiple representations. It is crucial for participants to flexibly use appropriate representations and shift between different representations in order to solve them effectively. Regarding item 19, it is expected to have algebraic and graphic representations of equation and inequality. With regard to item 21, it is necessary to shift between algebraic and graphic representations in order to solve the problem. To solve the problem of item 22, it is necessary to have ability in translating graphic representations to algebraic representations. To solve the problem of item 23, it is required to use appropriate forms of algebraic expressions and transformations of different algebraic expressions, and translate between graphic and algebraic representations.

Response to Item 19. There was a significant mean difference of this item between China and the U.S. (MD=2.90, *t*=34.62, *p*<.001). The score distribution of the item is displayed in Figure 4.12.

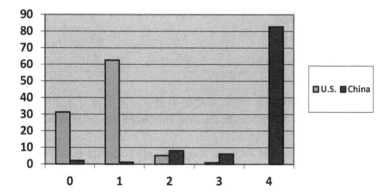

Fig. 4.12. Score distribution of item 19.

The figure showed that 82% Chinese participants gave two essentially different solutions to the inequality, while only one U.S. participant gave two essentially different solutions. Moreover, about 6% of Chinese participants gave two correct algebraic solutions while only one U.S. participant did the same. In addition, one-third (31.3%) of U.S. participants left it blank or gave some useless statements, while only 2.1% of Chinese participants did the same.

Strategies used in item 19. The different strategies or methods used to solve the inequality are displayed in Table 4.16. The Chinese participants demonstrated a high fluency and flexibility in solving the inequality. About 80% of participants from the high-achieving group and 51% of participants from low-achieving group gave two essentially different methods of solving the inequality. In addition, about 5% of the participants from each group gave two algebraic methods to solve the inequality. In contrast, the U.S. participants struggled with solving the quadratic inequality (only two participants gave two methods correctly).

Table 4.16. Different Methods of Solving Inequality in Item 19

Types of explanations	Frequency		
	China		U.S.
	High (%)	Low (%)	(%)
Two algebraic methods	5	6	0
Algebraic and interval sign	4	1	0
Algebraic and linear equation	1	1	1
Algebraic and quadratic function	80	51	1

Mistakes Made by the U.S. Participants in item 19. The U.S. participants revealed numerous misconceptions and mistakes as displayed in Table 4. 17.

The table shows that 44% of the U.S. participants adopted the inference: if $ab>0$, then $a>0$, $b>0$. None of them realized that a and b could be negative. In addition, none of them cared about the logical operations "or" or "and" between two logical propositions (such as $a>0$ and $b>0$ or $a>0$ or $b>0$). They also were satisfied with the solution "$x>3, x>-4$" without any intention to further intersect or combine the two.

Table 4. 17. Misconception or Mistakes Occurred in Solving Inequality in Item 19

Types mistakes	Explanation	Examples	Frequency (%)
1	Misconception: if $ab>0$, then $a>0$, $b>0$	$(x-3)(x+4)>0 \rightarrow$ then $x>3$, $x>-4$. $\quad x-3>0, \quad x+4>0,$	37
2	Only transforming into standard form	$x^2 + x - 12 > 0$ or $x^2 + x > 12$	21

Types mistakes	Explanation	Examples	Frequency (%)
3	Transforming into standard form and getting stuck	$x^2 + x - 12 > 0$ or x(x+1)>12 or x(x+1)=12	7.6
4	Working on the standard form with guess and check	$x^2 + x - 12 > 0 \rightarrow$ $x(x+1) > 12$; $\rightarrow$x>12, x+1>12$\rightarrow$ x>12, x>11	12
		$x^2 + x > 12 \rightarrow x^2 > x - 12 \rightarrow$ $x > \sqrt{x - 12}$	2.5
		$x^2 + x - 12 > 0$, or (x-3)(x+4)>0 $\rightarrow$x$_1$=3, x$_2$=-4.	15
5	Drawing number line	Find partial answer: x>3 or x<-4	4
6	Using a table	x>3 (x=1, 2, 3, 4... or 0, -1, -2, -3,...).	4
7	ab>0$\rightarrow$a>0/b or b>0/a	(x-3)(x+4)>0$\rightarrow$ x-3>0/(x+4), then x>3	1.6

In order to find another method of solving the inequality, an automatic alternative is to transform the factor form into standard form: $x^2 + x - 12 > 0$. 21% of them stopped with the standard form. 7.6 % of them were stuck with further algebraic operations: $x(x+1) > 12$ or $x(x+1) = 12$. Some of the participants went further with "guess and check strategies":

Mistake 1 (12%): $x^2 + x - 12 > 0 \rightarrow x(x+1) > 12$; $\rightarrow$x>12, x+1>12$\rightarrow$ x>12, x>11.

Mistake 2 (15%): $x^2 + x - 12 > 0$, or (x-3)(x+4)>0 $\rightarrow$x$_1$=3, x$_2$=-4.

Mistake 3 (2.5%): $x^2 + x > 12 \rightarrow x^2 > x - 12 \rightarrow x > \sqrt{x - 12}$

In addition, there were some unexpected mistakes as follows:

"(x-3)(x+4)>0$\rightarrow$ x-3>0, x+4>0, then x>3 & x>-4: -4<x<3";

" $x^2 + x > 12 \rightarrow x^2 > 12, x > 12 \rightarrow x > \sqrt{12}$ ";

"Solve by guess and check, x>3 because x=3 makes it zero, (x-3)

=(x+4), $3 \neq 4$, so x >3";

"(x-3)(x+4)>0, if x=3, then "(x-3)(x+4)=0, not greater than 0, so

(3, ∞)";

" $x^2 + x - 12 > 0$, $x^2 + x > 12$, x(x+1)>12, x>12 or x>11";

" $x^2 + x - 12 > 0$, $\sqrt{x^2} + x > \sqrt{12}$, 2x>$\sqrt{12}$ .x>$\sqrt{12}$ /2 "; and

" $x^2 + x - 12 > 0$, x^3>12."

Even though they were not able to find the correct answers, they took the risk of using different representations such as a number line, or table to explore the solution as follows:

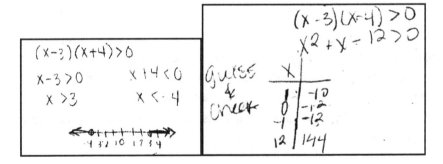

Figure 4.13. Different representations used in the U.S.

In summary, it is encouraging that the U.S. participants tried to use different representations to explore different ways of solving the inequality, and take the risk of "guess and check". However, it is disappointing that nobody attempted to use the graphing method, and almost all (except for one) were not able to provide a correct solution. Moreover, numerous misconceptions and mistakes were revealed when using the guess and check strategy.

U.S. participants' explanations to item 19. One of the interviewees (Larry) just simplified the factored form into standard form ($x^2 + x - 12 > 0$) and then got stuck on handling the solution. He moved forward by "guess and check" such as "$x^2 > x - 12$, and then $x > \sqrt{x - 12}$". He just square rooted them, even though it did not work (he knows that when dividing something in inequality, the in-equality sign may be changed, but she did not remember the exact rules).

By analogizing the property of equation: (x-3)(x+4)=0 → x-3=0, or x+4=0, the remaining four interviewees made the following an inference: $(x - 3)(x + 4) > 0 \rightarrow (x - 3) > 0$, $(x + 4) > 0 \rightarrow x > 3, x > -4$.

In order to find a second method, Kerri took a risk using guess and check:

$(x - 3)(x + 4) > 0 \rightarrow x^2 + x - 12 > 0 \rightarrow x(x + 1) > 12 \rightarrow x > 3$, while

Stacy used root formula $x_{1,2} = \dfrac{-1 + \sqrt{1 - 4(1)(-12)}}{2}$ =3 or -4, and got the same solution x>3, x>-4.

Nobody intended to work on "x>3, x>-4" in order to use logical operations such as "and" or "or" and the operations of intersection and combination of sets. They seemed to be satisfied with the "solution".

When asked "if *ab*>0, what result can you deduce?" they realized that "if *ab*>0, then *a,b* both are positive, *a, b* both are negative". Then they realized that there are other solutions to the inequality. Kerri used a number line to find a correct solution x>3 and x<-4. Two other (Alisa and Stacy) guessed that x<-4 should be part of solutions; however, they still worked with algebraic representation to solve this inequality.

When asked if they could use a graphic method to solve equation or inequality, they recalled the graphs of quadratic equations. Aside from Jenny (she drew one without intersecting with x-axis), the others drew a correct sketch and found the correct solutions with the support of the researcher. Moreover, Kerri not only presented the solutions by number line, but also drew two lines y=x-3 and y=x+4 to show how to use the graph of a linear equation to find the solutions of (x-3)(x+4)>0.

All the interviewees explained that they did not know how to use the graph of the quadratic function graph to solve the inequality, even though they knew the graphing method of solving a linear equation. They learned quadratic functions first (probably later at middle school or early at high school) and then quadratic inequalities later at high school. These contents were taught separately. They were not taught how to use graphic representations to solve algebraic problems. They appreciated the method of integration of algebraic and graphic representations. Stacy said "it will be better to teach students with two methods, because some students are visual learners while others are algebraic learners."

In summary, only one U.S. participant provided two correct algebraic solutions. Almost all of the U.S. participants struggled with algebraic computation of the inequality, with numerous mistakes when guessing and checking. However, the interviews revealed that participants might have had the relevant content knowledge (such as quadratic function and its graph, quadratic equation, and inequality), but they did not have an interconnected knowledge network; they did not have problem solving experience in flexible using different kinds of knowledge and relevant representations. However, when appropriately enlightened, they were able to build the connection between different types of relevant knowledge and find an appropriate solution. In addition, as pointed out by the interviewees, the placement and presentation of the contents in textbooks and the ways of teaching the contents in classroom did not support in building the connections between algebraic and graphic representations. This raises an important issue related to teachers' knowledge development.

Chinese participants' strategies used in item 19. On the other hand, the Chinese participants provided multiple methods to solve the inequality. Four-fifths of the participants provided two essentially different solutions (one is using algebraic manipulation and the other is using a graphing method). They provided correct procedural steps and did not make basic mistakes similar to those made by the U.S. participants. For example, one participant gave the following two typical solutions:

Method 1: since (x-3)(x+4)>0, then (x-3)>0 and (x+4)>0 $\Rightarrow$,x>3

and x>-4 or x-3<0 and x+4<0, $\Rightarrow$ x<3 and x<-4. So

the solution is $\{x \mid x > 3 \text{ or } x < -4 \}$;

Method 2: (graphing method): According to the graph of function

f(x)=$(x-3)(x+4) = x^2 + x - 12$, when

$x \in (-4,3), f(x) < 0.$ when $x \in (-\infty,-4) \cup (3,\infty), f(x) > 0.$

Thus, the solution of the inequality is $(-\infty,-4) \cup (3,\infty)$.

One student gave a graphing method as follows: Sketch two lines: y=x-3 and y=x+4, then find the common regions where both lines are positive (above the x-axis) or negative (below the x-axis). The x intervals of those regions are the solutions.

Summary of responses to item 19. The above analysis shows that Chinese participants had sound knowledge in solving the inequality both algebraically and graphically. However, the U.S. participants lacked relevant knowledge and skills in solving inequalities by using graphic methods, and made a lot of basic mistakes when trying to find solutions. They did not realize that they could solve the inequality by quadratic function graphs.

Response to Item 21. Item 21 is used to measure teaching and student knowledge of solving quadratic equations. It is necessary to link algebraic and graphic representations. There was a significant mean difference of this item between Chinese and the U.S. participants (mean difference=2.79, t=31.60, p<.001). The score distribution of this item is displayed in Fig. 4.14.

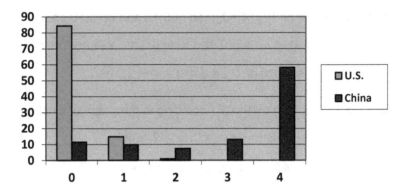

Fig. 4.14. Score distribution of item 21

Of the U.S. participants, only one gave correct explanations and useful suggestions as shown below.

(a) The student believes that he needs to know the values of a, b, c. He can't find these values because there are three variables in two equations.

(b) The student needs to think graphically. Since at x=1, the value is positive, then the graph is above the x-axis. The opposite is true when x=6. Therefore, the graph has to cross the x-axis and since it has degree two. It must have 2 solutions.

About 84% of them agreed with this student's explanation (even though it is wrong), and they were stuck with the algebraic operation to find a, b and c, and had no idea how to help the student. 15% of the participants suggested plugging different values of a, b, and c (such as a=-10, b=-9, and c= 20) to see whether a pattern could be found. It is disappointing that the U.S. participants have no idea how to think with graphic or geometrical representations; they even faced difficulty using algebraic representation.

Among the Chinese participants, about three-fifths (58.2%) of the participants gave appropriate explanations of students' mistakes and provided a correct answer. For example, one participant gave detailed explanations of the student's reasons for his/her solution:

"Reasons: First, although the student masters some methods of solving problems, she/he directly applies that knowledge without considering the specific conditions of the problem. He/she was constrained by routine thinking methods; second, the student did not recall the method of judging zero points of function. The method can be used flexibly.

Suggestions:

(1) if you are not able to find a, b, and c, by using the given conditions (when $x=1$, the value is positive while $x=6$, the value is negative), why not try other methods ? Are there any ways which do not require to find a, b and c?

(2) Although we are not able to find accurate values (of roots) by using algebraic methods, then, why not use graphing method to make estimations? It should be helpful to guide students to draw a figure (similar to the Figure 4.13)

(3) Through observing this figure, are there any intersection points of the function at x-axis?

(4) How many intersection points are there? According to the features of quadratic function, students could solve this question?

(5) Connecting zero points of a function to the roots of an equation lets students understand that learned knowledge can be flexibly applied to solve this problem. "

For another example, one participant gave a solution (See the Fig. 4.15).

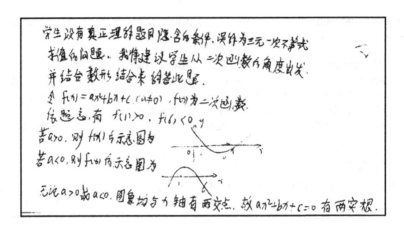

Fig. 4.15. An example of Chinese answers to item 22.

The participant believed that "the student did not fully understand the hidden condition of the problem and mistreated it as a problem of finding solution of a quadratic inequality". She/he further suggested the student is to consider the problem by integration of algebra and geometry from a perspective of quadratic function. Then s/he drafted two sketches of the quadratic function ($f(x) = ax^2 + bx + c$, a>0 or a<0) according to the given conditions $f(1) > 0, f(6) < 0$, and concluded that there are two roots of the quadratic equation.

Another 13% of the participants identified students' problems and gave correct answers with minor computational or notational errors. Another 7% of the participants realized students' mistakes and suggested using a graphing method, but without any details. About 10 % of them made efforts to find a, b and c or determine the sign of the discriminant. The remaining 10% left the item blank or wrote something not useful for solving the problem.

However, some participants tried to determine the number of roots based on the sign of the discriminant ($\Delta = b^2 - 4ac$), and then they got stuck in some inappropriate algebraic manipulations. For example, one participant gave the following explanation:

Solution: if x=1, a+b+c>0; if x=6, 36a+6b+c<0, so 35a+5b<0; 7a+b<0, b<-7a.

(1) If b>0, c>0, then, a<0, then $b^2 - 4ac > 0$, then the equation has two roots.

(2) If b>0, c<0, then, a<0...

It is important to learn how to discuss and solve a problem according to different parameters, based on the sign of $\Delta \geq 0$ (two different roots, no real roots, and two equal roots). Since there are several parameters, usually, we fix the values of some parameters, and then adjust other parameters. Thus, the discussion will be very clear. (SC04-28).

Strategies used in item 21. The interpretations used in solving item 21 are displayed in Table 4.18

Table 4.18. Interpretations Used in Item 21

Interpretations	Frequency		
	China		U.S. (%)
	High (%)	Low (%)	
Correct explanations and correct graphing solutions	76	65	1
Using graphing method in general without providing solutions in detail	4	8	0

As analyzed above, the Chinese participants demonstrated an ability to use graphic methods to solve the algebraic equation. Moreover, compared with the low-achieving group, the high-achieving group seems to provide more complete and detailed solutions using the graphing method.

In Even's (1998) study, it was found that only 14% of the 152 prospective secondary mathematics teachers in the U.S. correctly solved this problem, and about 80% of them did not show any attempt to look at another representation of the problem. In the current study, only 1% of the 115 U. S. gave a correct answer while around 58% of 376 Chinese counterparts gave fully correct answers. The U. S. subjects in this study performed very poorly, and it may be due to the fact that 80% of the U. S. participants were preparing to be middle school math and science teachers. The accuracy of the Chinese prospective teachers is higher than that (14%) of prospective mathematics teachers in Even's (1998) study. Thus, the Chinese participants demonstrated strong knowledge and skills in shifting between symbolic and graphic representations when solving this problem.

U. S. Participants' explanations to item 21. Three of the participants (Larry, Kerri, and Alisa) fully agreed with the student's statement. Namely, "Since it is impossible to find out fixed values of *a, b* and *c* based on the previously given inequalities, the original question is not solvable". They tried to find out *a,b,* and *c* through algebraic transformation but it did not work. They had no idea how to help the student find a solution.

The other two interviewees (Jenny and Stacy) felt the problem could be solved, but they did not have any concrete ideas on how to solve it. What they could suggest to the student was to "try different ways, such as plugging more numbers between 1 and 6."(Jenny) or "explore in different ways such as plugging more numbers to see whether they can find certain patterns, rather than being stuck" (Stacy).

When asked whether they could try other methods such as graphical methods to solve, they tried to sketch the graphs and find the possible roots. Except for Larry, others were successful in finding the number of roots by examining the intersection points of the quadratic function. All of them said they had not thought in this way. They had not participated in these kinds of experiences in solving problems, but they realized the usefulness of graphing method in algebra.

Summary of responses to item 21. The Chinese participants (more than 60%) demonstrated flexibility in using graphic representations to solve this problem, and only a small part of them (20%) were stuck with algebraic operations. However, of the U.S. counterparts, 85% struggled with algebraic manipulations, and failed to find correct answers and explain the students' mistakes. Less flexibility in using graphic representations was revealed when finding the number of roots of the quadratic equations.

Response to Item 22. This item is used to measure knowledge for understanding of the effects of changes of parameters of quadratic function on the changes of quadratic graphs. There was a significant mean difference between China and the U.S. (MD=1.42, t=-8.44, p<0.01). The score distribution of the item is displayed in figure 4.16.

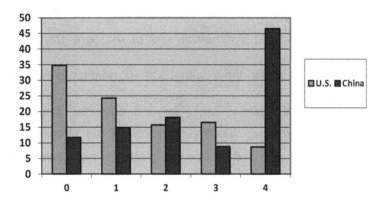

Fig. 4.16. Score distribution of item 22.

About 10% of U.S. participants gave correct answers and appropriate explanations and 17% of them gave correct answers but failed to explain. Fifteen percent of the participants gave partially correct answers and explanations. Twenty three percent of them gave sporadic information about the effect of *a, b* and *c* changes. About 35% of them got lost, either leaving it blank or providing some wrong statements. In summary, one-third of U.S. participants had no ideas how to solve and explain this problem while about one-fourth of them gave somewhat correct answers.

The graph also showed that 47% of Chinese participants gave the correct choice and appropriate explanations. They either explained by analyzing the effects of changes of *a, b,* and *c* on the changes of the graphs or analyzed the symmetrical features. For example, the following are excerpts:

"Because the change a results in changes of graph in the openness and wideness, however, the translated graph does not change the shape, so a is not changed. Since c represents the y-intercept, because the two graphs intersect y- axis at the same point. So c is not changed. So, only b can be changed." (Method 1, SC04019)

"According to the graphs, a>0, the two graphs are symmetrical with regard to y-axis, thus, the two symmetrical lines are also symmetrical with regard to y-axis. The symmetrical line of the left graph is $x = -\dfrac{b}{2a}$, so the symmetrical line of the right graph should be $x = \dfrac{b}{2a}$. Thus, only the b changed as –b" (Method 2, SC409).

"Let original form: $y = ax^2 + bx + c$, and the translated one: $y_1 = a_1 x^2 + b_1 x + c_1$. Since y-intercepts are the same, namely, x=0, y=c= $y_1 = c_1$, so, c= c_1. Since the graph y1 and y are symmetrical with regard to y- axis, so:

$$y = ax^2 + bx + c \Rightarrow y_1 = a_1(-x)^2 + b_1(-x) + c_1$$

$$ax^2 + bx \Rightarrow a_1 x^2 - b x_1 \Rightarrow a = a_1, b = -b_1$$. Thus, it is only needed

to change b value." (Method 3, SC411)

Another 9 % of the participants gave a correct choice, but their explanations included minor errors. About 18% of the participants made correct choices and their explanations included serious mistakes.

Another 15% gave wrong choices but provide some useful explanations, and the remaining 12% just gave up any efforts to answer the item.

The strategies used or mistakes made in item 22. The following table (Table 4.19) showed the different interpretations used.

Table 4.19. Interpretations Used in Item 22

Interpretations	Frequency		
	China		U.S.
	High (%)	Low (%)	(%)
The effects of changes of a, b and c on the changes of graphs of quadratic functions. (Method 1)	7	14	11
Symmetrical line is $x = -\dfrac{b}{2a}$, a is invariant, then b can be changed only (Method 2).	31	27	2
According to g(x) =f(-x), find the coefficients of g(x) ($a_1 = a$, $b_1 = -b$, $c_1 = c$). (Method 3)	11	14	0

The table showed that the Chinese participants used more ways (see previous explanation for details of the three methods) to interpret their answers than their U.S. participants did (two methods). Chinese participants not only provided the general discussion on the effects of changing parameters on the changes of graphs (method 1), but also used the properties of symmetry both geometrically (method 2) and algebraically (method 3). However, the U.S. participants mainly used method 1 which was described in typical texts.

Meanwhile, since the Chinese participants used more sophisticated algebraic multiplication, they made some errors. For example, one participant tried to transform the function expression as follows:

Because the translated function should be: $g(x) = f(x - k) =$

$$a(x - k)^2 + b(x - k) + c = ax^2 - (bk + 2ak)x + (ak^2 - bk + c).$$

Thus $ak^2 - bk + c = c$, $ak^2 - bk = 0$, k=0 or $k = \dfrac{b}{a}$. So at least two

of these parameters of a, b, c should be changed.(SC04-22)

Similarly, another participant tried to find the vertex point: $x_1 = \dfrac{b}{2a}$,

$y_1 = \dfrac{4ac - b^2}{4a}$, and explained that " since the y1 is the same [at the two

vertex points of the two graphs], so only x1 could be changed. Thus, at least two parameters need to be changed." (SC-04-24).

In Black's (2008) study, 20% of 76 U.S. high school mathematics teachers gave correct answers and relevant explanations to this problem. In the current study, 25% of 115 U.S. participants gave correct answers and explanations while 55% of 376 Chinese counterparts did the same. In this item, the U.S. participants in this study performed better than the subjects in Black's study. The Chinese participants in this study outperformed the U.S. counterparts remarkably.

U.S. interview participants' explanations to item 22. In the interview, two participants (Larry, and Kerri) clearly explained the effects of changing *a, b*, and *c* on the graphs of quadratic functions, even though Kerri made a wrong choice in the survey.

Alisa and Stacy were able to explain the effect of changing *a* and *c* on the graph of quadratic function, but they were not sure about the effect of changes in b. Alisa made mistakes in drawing graph $f(x + h)$. Stacy knew how changes of *a* and *c* impact the changes of the graph but she was not clear about how changing *b* can impact the changes of graphs although she got a correct choice.

Jenny found the correct answer by explaining that changing b to negative b, the graph of quadratic function would reflect it over the y-axis because she "did a lot of exercises of translation of graphs in high school". However, she could not remember the details of the effect of changing *a, b* and *c* on the graph.

In summary, two of the interviewees were quite clear about how the changes of a, b impact on the changes of the graphs of quadratic function. Others were not quite sure how changes of these parameters affect the changes of the graphs of quadratic function. One interviewee got the correct answer by relating to the symmetric property, although she was not clear about the details of the effects of changing *a, b* and *c* on the graph.

Summary of the responses to item 22. More than half of the Chinese participants provided the correct choice and roughly appropriate explanations (46.5%+8.8%=55.3%) while only about one fourth (10%+17%=27%) of the U.S. counterparts did the same. In contrast, about one tenth (11%) of Chinese participants gave up any attempt to solve the problem, while more than one-third (35%) of the U.S. participants gave up. In addition, Chinese participants attempted in various ways to interpret using the connections between geometrical symmetry properties and algebraic function features, while their U.S. counterparts mainly interpreted in the typical way introduced in textbooks.

Response to item 23. There was a significant mean difference of item 23 between China and the U.S. (MD=3.00, t=33.01, p<.001). The score distribution of the item is displayed in figure 4.17.

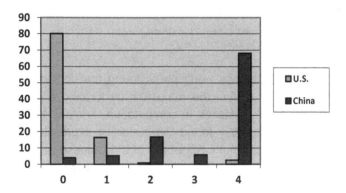

Fig. 4.17. Score Distribution of Item 23.

Only three U.S. participants found the quadratic equation by solving a system of linear equations and found the maximum correctly. One participant found the quadratic equation but failed to find the maximum. About 20% of the participants just drew a graph or listed equations based on the given three points. Four-fifths of them left the item blank or wrote some unrelated statements.

The figure shows that 68% of the Chinese participants correctly solved this problem. Typically, they used standard form of a quadratic (i.e., $y = ax^2 + bx + c$) to find the function expressions, and then transformed it into a vertex form (i.e., $y = a(x - h)^2 + k$) to find the maximum. However, multiple strategies were used to find the solutions (see Fig 4.18). For example, one method used the factored form of a quadratic: $y = a(x - x_1)(x - x_2)$, while the other method used the standard form $y = ax^2 + bx + c$ and Vièta's theorem: $x_1.x_2 = \dfrac{c}{a}$, $x_1 + x_2 = -\dfrac{b}{a}$ (a=-2).

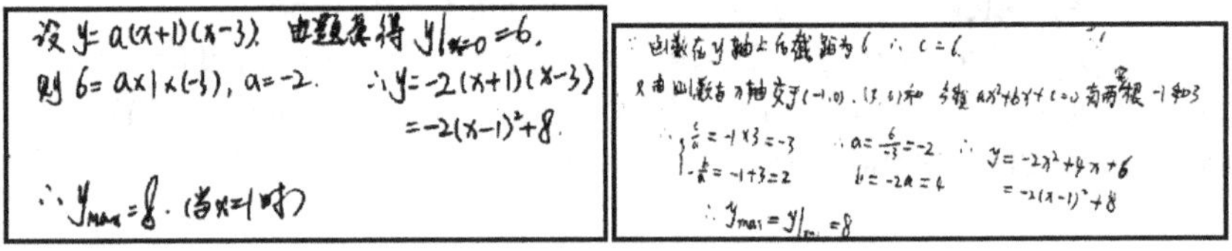

Fig. 4.18. Two methods used in item 23 in China.

About 6% of the Chinese participants found the correct quadratic expressions and tried to find the maximum by forming a perfect square, however, they made mistakes in computation. Another 17% of the Chinese participants correctly found the quadratic expression, without further action to find maximum. About 5% of the Chinese participants could only find part of a, b and c, but failed to find the expression, and the remaining 4% just left the item blank.

Strategies used in item 23. With regard to the strategies used to solve this problem, there are several methods. First of all, three forms of a quadratic function were used: $y = ax^2 + bx + c$ (FM1); $y = a(x - x_1)(x - x_2)$ (FM2); and $y = a(x - h)^2 + k$ (FM3) could be used to find the quadratic function expression. Then, three methods could be used for finding the maximum value: (1) transforming into $y = a(x - h)^2 + k$, then finding the maximum (MM1); (2) using formula $x = -\dfrac{b}{2a}$, $y_{max\,imum} = \dfrac{4ac - b^2}{4a}$ (MM2); and (3) taking derivative: $\dfrac{dy}{dx} = 0$

to find x=1, then, $y_{maximum} = f(1)$ (MM3). The strategies used are shown in Table 4.20.

Table 4.20. Different Strategies in Using Representations in Item 23

Strategies	Frequency		
	China		U.S.
	High (%)	Low (%)	(%)
1.FM1+MM1	22	18	3
2.FM1+MM2	28	32	0
3.FM2/FM3+MM1	12	1	0
4.FM2/FM3+MM2	8	10	0
5.FM1/FM2/FM3+MM3	4	1	0

The table shows that the Chinese participants provided various strategies to solve the problem. The high achieving group showed more variability in adapting strategies (five methods) than did the low achieving group (essentially including three methods). Moreover, some participants (20% in high achieving group and 11% in low achieving group) demonstrated an ability to select the most appropriate format (strategies 3 or 4). The following are some examples of flexibly using formula and properties of quadratic functions. For example, a Chinese participant from the high achieving group gave two methods:

Method 1: using standard formula to find the quadratic function and completing the square to find the maximum (i.e. strategy 1).

Method 2: according to the given condition, the symmetric line is x=1. Let $f(x) = a(x - 1)^2 + h$.

Because f (-1)=0, f(0)=6.

So, 4a+h=0 and a+h=6 $\begin{cases} 4a + h = 0 \\ a + h = 6 \end{cases}$; $\begin{cases} a = -2 \\ h = 8 \end{cases}$, thus,

$f(x) = -2(x - 1)^2 + 8$. Thus the maximum is 8. (i.e. strategy 3)
(SC04-15)

Another participant from the low achieving group also gave two methods as follows:

"Solution: c=6.

Method 1: Let $ax^2 + bx + c = 0$. Since $x_1 = -1, x_2 = 3$ are two roots, c=6, so,

$x_1.x_2 = \dfrac{c}{a}$, a=-2; $x_1 + x_2 = -\dfrac{b}{a}$, b=4

Thus, f(x)= $-2x^2 + 4x + 6$, and symmetric line is x= $\dfrac{x_1 + x_2}{2} = 1$,

$F(x)_{max} = f(1) = 8$.

Method 2: Let y=a$(x - 3)(x + 1)$, plugging x=0, y=6, find a =-2, plugging the symmetric line x-coordinate (x=1) and find the maximum."
(LN4-02)

It is impressive that the participants flexibly used Vièta's theorem in method 1 and wisely chose an appropriate formula of the quadratic function in method 2.

The U.S. participants revealed a lack of basic skills of algebraic computation. In the following section, we try to get a better understanding of U.S. participants' thoughts about solving this problem.

U. S. participants' explanations of their solution. Two of the participants (Larry and Stacy) knew the process of solving the problem: finding the expression of quadratic equations by plugging given points, and then finding the maximum by taking the derivative. Two of them (Larry and Alisa) realized that the maximum should be at x=1 due to its symmetry although they were not able to find correct expression of the quadratic equation. Two of them (Kerri and Jenny) supposed that the y-intercept is the maximum. They had difficulties finding the expression by plugging in the given points.

When asked whether they could use another form of quadratic equations to find an expression more effectively, they had no idea. Even when I showed them some formulas (such as $y = a(x - x_1)(x - x_2)$ or $y = a(x - h)^2 + k$), they did not know about them.

In summary, the prospective teachers tried to find expressions of quadratic equations by using the standard formula, and then found the maximum by taking the derivative and plugging x=1. However, they had difficulty finding the correct expression due to the complexity of algebraic manipulation. Some of them tried to use the symmetry to find maximum. In addition, nobody was aware of using other appropriate form to find the expressions as Chinese counterparts did. It seems that the standard form of a quadratic equation is the only one they were familiar with.

Summary of the responses to item 23. The analysis of the responses to the item 23 showed that Chinese participants demonstrated a sound and flexible knowledge for solving this problem. They not only could shift from process perspective (point-wise) to object perspective (graphic and algebraic representations) but also flexibly selected the most appropriate form (algebraic representations) to solve this problem. On the other hand, the U.S. participants revealed a shortage of using relevant knowledge to solve the problem.

Summary of the Representational Flexibility in China and the U.S. The analysis of the responses to the four items which focused on solving and interpreting quadratic functions/equations/inequalities provides a consistent picture of teacher knowledge for teaching the concept of quadratic relation. Overall, the Chinese participants not only demonstrated a sound knowledge needed for teaching the concept, but also showed the flexibility in using representations appropriately. In contrast, the U.S. counterparts revealed their weakness in basic knowledge for teaching the concept, and in flexibility in using representations.

4.4   An Analysis of Correlation between Flexibility and Other Variables

Due to the importance of developing teachers' flexibility in using appropriate representations, I developed an indicator of flexibility. Flexibility in this study is defined as a shift between different representations or a transformation between different forms within a representation.

For example, in item 22, if a participant explained as follows: "since the symmetrical line is $x = -\dfrac{b}{2a}$ while the openness of the graph is not changed so the parameter a will not be changed, thus, only b can be changed", I coded the response as a flexibility (shift from graphic representation to algebraic representation). So, each of the three strategies used (see table 4.19) can be coded as a flexibility.

In another example, in item 23, if a participant adopted the standard form of a quadratic function (i.e., $y = ax^2 + bx + c$) to find the expressions and then reorganize it as a form of $y = a(x - h)^2 + k$ in order to find its maximum, I coded this event a flexibility (transformation within algebraic representation). Thus, each of the strategies 1-4 used for solving item 23 (see Table 4.20) was coded as a flexibility.

The number of times that flexibility occurred in items (19, 21, 22, & 23) were counted as the value of a variable denoted as flexibility, ranging from 0 to 4. The indicator of flexibility was examined through two aspects: (1) multiple group comparison; and (2) regression analysis with regard to SM, AM, and TM as independent variables in different groups.

Mean difference across different groups. The means scores and standard deviation of flexibility are displayed in Table 4.21.

Table 4.21. Mean Scores and Standard Deviation of Flexibility

	Mean	St. Deviation
1.China High-achieving	3.09	1.050
2.China Low-achieving	2.76	1.190
3.U. S.	.29	.491

Multiple group comparison shows that the mean difference between Chinese high-achieving and low-achieving group is significant (mean difference=.33, p=0.008). The mean difference between Chinese low-achieving group and the U.S. group is also significant (mean difference=2.47, p=0.000). Thus, there are significant differences of flexibility between different groups of participants. The higher KAT score the participants have, the more flexible they are in selecting representations.

Prediction of different components of KAT. Using flexibility as the dependent variable and three components of KAT as independent variables, I did a regression analysis in different groups. (let y=Flexibility, x_1=School mathematics, x_2=Teaching mathematics, x_3=Advanced mathematics)

In the Chinese high-achieving group, the regression equation is as follows:

$$y = -1.09 + 0.12x_1 + 0.17x_2$$

Knowledge of school mathematics and teaching mathematics explains 56% of the variance in flexibility ($R^2 = 0.56$, F (2)=67.16, p<0.001).

In the Chinese low-achieving group, the regression equation is as follows:

$$y = -1.69 + 0.26x_1 + 0.12x_2$$

Knowledge of school mathematics and teaching mathematics explain 74% of variance in flexibility ($R^2 = 0.74$, $F(2)$=200.9, p<0.001).

In the United States, the regression equation is as follows:

$$y = -.50 + 0.03x_1 + 0.09x_2$$

Knowledge of school mathematics and teaching mathematics explain 42% of variance in flexibility ($R^2 = 0.42$, $F(2)$=40.80, p.<0.001).

However, in the whole sample (including Chinese and U.S. participants N=371), the regression equation is as follows:

$$y = -1.41 + 0.19x_1 + 0.12x_2 + 0.03x_3 \ .$$

Knowledge of school mathematics, teaching mathematics and advanced mathematics explain 84% of variance in flexibility ($R^2 = 0.84$, $F(3)$= 663.84, $p.<0.001$).

At the same time, flexibility is highly correlated to each of school mathematics(r=.87, p<.001), teaching mathematics (r=.86, p<.001) and advanced mathematics (r=.81, p<.001).

In summary, overall, the higher KAT scores the participants achieve, the higher flexibility they have. Meanwhile, the flexibility can be significantly predicted by KAT and it is highly correlated to all the components of KAT.

4.5   Summary of the Findings

The findings of this study can be summarized in line with four research questions: (1) the differences and similarities of KAT in Chinese and U.S. prospective teachers, (2) the relationship between different components of KAT, (3) the difference and similarities of knowledge for teaching the concept of functions, and (4) the relationship between prospective teachers' status of KAT and their courses taken.

4.5.1 The Differences and Similarities of KAT in Chinese and U.S. Prospective Teachers

In all 17 multiple choice items, except for four items, the Chinese participants had significantly higher mean scores than their U.S. counterparts. In all 8 open-ended items, the Chinese participants significantly outperformed their U.S. counterparts. Based on detailed examination of individual items, I found that: (1) the U. S. participants showed a better understanding of introducing the slope concept from multiple perspectives than the Chinese counterparts; (2) both the U. S. and Chinese participants revealed weaknesses in presenting numerical relations and algebraic equations using geometrical representations; (3) the Chinese participants tended to make their judgments based on visual and graphical information and underlying conceptual understanding and logical reasoning while their U. S. counterparts tended to make their judgments mainly based on visual information without paying close attention to underlying concepts and logical reasoning; and (4) the Chinese participants demonstrated strong knowledge and skills in algebraic manipulation and quadratic functions/equations/inequalities.

4.5.2 The Relationship between Different Components of KAT

With regard to the U.S. sample, school mathematics was found to have significant effects on teaching mathematics and advanced math. Advanced math was found to have a direct and significant effect on teaching math. Regarding the Chinese sample, school mathematics was found to have direct and significant effects on advanced math and teaching math.

4.5.3 Difference and Similarities of Knowledge for Teaching the Concept of functions

The analysis of the items of measuring teachers' knowledge for teaching the concept of function revealed the following results: (1) the Chinese participants demonstrated sound knowledge and skills needed for solving the problems and interpreting their solutions, while their U.S. counterparts revealed their limitations in basic knowledge and skills; (2) the Chinese participants showed flexibility in selecting appropriate perspectives of the function concept while the U.S. counterparts showed disadvantages in adopting appropriate perspectives; and (3) the Chinese participants demonstrated flexibility in using multiple representations while the U.S. counterparts revealed limited knowledge and ability in adopting multiple representations appropriately. In addition, the Chinese participants were willing to provide more diverse interpretations than their U.S. counterparts.

There are significant differences of flexibility between China and the U.S. The Chinese participants demonstrated greater flexibility in using representations and perspectives than their U.S. counterparts. Overall, the KAT scores are highly correlated to the flexibility and the flexibility can be significantly predicted by KAT.

4.5.4 The Relationship between KAT and Courses Taken

In China, the number of math courses taken was found to have a significant effect on advanced mathematics, but the effects of the number of courses taken on school mathematics and teaching mathematics were not significant. In the U.S., the number of math courses taken was found to have significant effects on school mathematics and teaching mathematics, but it did not have significant effect on advanced mathematics. These findings imply that the Chinese teacher preparation programs may emphasize content knowledge while the U.S. teacher preparation programs may emphasize pedagogical knowledge.

5 Chapter Five: Conclusion and Discussion

Before discussing my findings, I would like to point out the disparity of sampling and courses taken between China and the U.S. First, the U.S. sample was a convenience sample taken from an interdisciplinary middle grade math and science program in a large public university. The U.S. participants came primarily from one of three routes of middle grade mathematics teacher preparation: a program designed exclusively for middle grade teachers' preparation. The Chinese participants were sampled from the secondary prospective mathematics preparation programs. In China, secondary school includes middle and high schools. There is only one type of secondary mathematics teacher preparation program (including middle and high schools) although there are variations regarding course design and arrangement. I purposely selected sample universities (high, normal, and low reputation universities). On average, the U.S. participants had taken seven mathematics content and mathematics education courses, while the Chinese participants had taken 14 such courses. Caution should be taken in interpreting the findings of this study in which comparisons are made between Chinese and U.S. participants due to the disparity of the sample.

In this chapter, I first summarized and discussed the main findings of this study in line with the research questions. Then, I discussed the limitation of this study and proposed some topics for further studies.

5.1 Conclusion

5.1.1 Knowledge of Algebra for Teaching in China and the U.S.

The data analysis showed overall the Chinese participants had a much greater knowledge of algebra for teaching than their U.S. counterparts. When looking at the items in detail, several interesting results were found as follows: (1) the U.S. participants showed a better understanding of introducing the concept of slope from multiple perspectives than the Chinese counterparts; (2) both the U.S. and Chinese participants revealed weaknesses in presenting numerical relations and algebraic equations using geometrical representations; (3) the Chinese participants tended to make their judgments based on visual information and underlying conceptual understanding and logical reasoning while the U.S. participants tended to make their judgments mainly based on visual information without paying close attention to underlying concepts and logical reasoning; and (4) Chinese participants demonstrated strong knowledge and skills in algebraic manipulation and quadratic functions/equations/inequalities.

Parts of these findings extend Ma's (1999) findings that Chinese elementary mathematics teachers had a profound understanding of fundamental mathematics knowledge to include middle school teachers. Moreover, Li and his colleagues (2008) indicated that the secondary (including middle grade) mathematics preparation programs in China put great efforts into developing students' sound and broad content knowledge and mathematics education knowledge. Thus, it is reasonable to expect that Chinese prospective secondary mathematics teachers may have a solid foundation of mathematics content knowledge. On the other hand, it is a publicly acknowledged observation that many U.S. mathematics teachers do not have adequate mathematics knowledge that they need to teach (e.g., Ball & Bass, 2000; CBMS, 2001). The study further alerted that the U.S. prospective middle grade teachers need to make great efforts to meet the recommendations of influential documents (CBMS, 2001; Kilpatrick et al., 2001; NMAP, 2008). The weaknesses of the U.S. middle grade mathematics preparation programs were identified by international comparative studies (Babcock et al., 2010; Schmidt et al., 2007; Tatto et al., 2011). The teacher preparation programs in East Asia including Korea and Chinese Taiwan demonstrated their strengths in mathematics content knowledge and pedagogical content knowledge compared with the U.S. programs. This study suggests that the middle grade mathematics teacher preparation programs in China share some features with other East Asian teacher preparation systems such as emphasizing mathematics content knowledge and pedagogical content knowledge, rather than pedagogical knowledge in general.

The weakness in presenting numerical relations and algebraic equations using geometrical representations in China and the U.S. calls for preparing teachers with connections of different knowledge and flexibilities in using different representations because they need to implement mathematics curricula that require such connection and flexibility (CCSSI, 2010; NCTM, 2000, 2009). It was recommended that students should understand the meaning of equivalent forms of expressions, equations, inequalities, and relations (NCTM, 2000), the connection between algebra and geometry, the link of expressions and functions, and the flexible use of representations (CCSSI, 2010; NCTM, 2009). Thus, prospective teachers should be equipped with relevant knowledge and skills so that they might be able to organize their classroom instruction with the necessary learning opportunities.

The observation that Chinese participants tended to make their judgments based on underlying conceptual understanding and logical reasoning while the U. S. tended to make their judgments mainly based on visual information may partially echo Cai's (2005) finding that Chinese teachers put more value on abstract representation than U.S. counterparts. The tendency of using abstract and logical reasoning of Mainland Chinese teachers was supported by Huang and Leung (2004)'s study. Their study comparing the ways of justifying Pythagoras's theorem revealed that Mainland Chinese teachers preferred to use mathematical proofs with algebraic manipulation rather than using visual verification, as Hong Kong teachers did.

Prospective teachers may be able to bring their previous learning experience into their reasoning and decision-making (Ball, 1990). Some studies found that Chinese students preferred to use abstract representations (Cai, 1995), and performed better on tasks that required no-visual representations (Brenner, Herman, Ho, & Zimmer, 1999), compared with U.S. counterparts. Furthermore, Chinese mathematics teaching is well known by its emphasis on mastering mathematics knowledge and skills and rigorous mathematics reasoning (Huang & Li, 2009; Leung, 1995, 2005; Li & Huang, 2013). It is plausible that Chinese prospective teachers preferred to make their judgments based on underlying concepts and logical reasoning, rather than just on the visual information. On the other hand, mathematics teaching in the U.S. classroom has always been described as emphasizing low-level, rather than high-level cognitive processes (i.e., memorizing and recalling facts and procedures, rather than reasoning about and connecting ideas or solving complex problems) (Hiebert et al., 2005; Silver, Mesa, Moriss, Star, & Benken, 2009; Wood, Shin, & Doan, 2006). If taking the U.S. students' preference in using visual representations and lack of ability in mathematical reasoning are taken together, then it may be understandable why U.S. prospective teachers tended to make their justification based on visual information given, without paying close attention to underlying concepts.

5.1.2 The Relationship between Different Components of KAT

The path analysis revealed that components of KAT in the Chinese sample are much more highly correlated than those in the U.S. sample. This implies that Chinese participants have a more interconnected KAT structure than U.S. counterparts. Measurement model analysis also confirms that Chinese participants have a highly correlated KAT structure.

Prospective teachers' knowledge structure is mainly impacted by the ways that they have been taught in both pre-university and university. Cai and Wang (2010) found that Chinese expert teachers put more emphasis on the coherence of delivering a good lesson than their U.S. counterparts. Moreover, Huang, Li, and He (2010) revealed that both novice and expert teachers in China viewed coherently developing a lesson as one salient feature of effective teaching. When examining classroom instruction, coherence and connection of lessons are essential features of teaching in China (Chen & Li, 2010; Wang & Murphy, 2004). Lesson coherence could be carried out through organizing systematic and varying classroom activities by sticking to mandatory well-designed textbooks (Huang, Ozel, Z. E., Li, & Rowntree, 2013; Huang, Li, & Ma, 2010). At the university level, lecture is the dominating teaching method (Li, Zhao et al., 2008), which may be conducive to transmitting knowledge systematically. It is possible that prospective teachers develop their well-structured and interconnected knowledge bases through their previous learning experience.

5.1.3 The Difference and Similarities of Knowledge for Teaching the Concept of Functions

The open-ended items were used to measure teachers' knowledge for teaching the concept of function in terms of the perspectives adopted and representations used. In general, a function concept should be developed from process to object perspective (Briedenbach et al., 1992; Sfard, 1990, 1993). A deep understanding of a function concept could be partially demonstrated by the flexibility in taking an appropriate perspective or shifting between these two perspectives. Moreover, using multiple representations and shifting between different representations are the manifestations of understanding a function concept and relevant skills. The analysis of the open-ended items from multiple aspects showed that the Chinese participants seemed to have:

(1) Strong algebraic and graphic transformational skills and procedural fluency;

(2) Multiple strategies of solving algebraic problems by integrating algebraic and geometrical representations;

(3) Appropriately taking perspectives and shifting between different perspectives of function; and

(4) Appropriate use of representations and flexible shifts between different representations.

On the other hand, the U.S. counterparts struggled with basic algebra manipulation and had limited knowledge of flexibly using perspectives and representations. Both qualitative and quantitative analyses showed that Chinese participants seemed to be more flexible than U.S. counterparts in terms of shifting between different perspectives and selecting appropriate representations of function.

It was found that the scores of KAT predict flexibility significantly, and the scores of KAT can explain more than four-fifths of the variance of flexibility. Meanwhile, the number of college math courses taken and grade level were not found to have significant prediction of the growth of flexibility. These results imply that developing flexibility is not a methodology and/or maturation issue, rather a comprehensive issue with the development of knowledge for teaching. This implies that prospective teachers need to be equipped with a well-structured knowledge base in order to develop their flexibility.

How can Chinese prospective secondary school teachers develop their flexibility while developing their procedural fluency and conceptual understanding? Many studies explored how Chinese in-service teachers develop their professional knowledge and expertise (Ma, 1999; Huang & Bao, 2006; Huang & Li, 2009; Li, 2004; Li &Li, 2009; Li, Huang, & Yang, 2010; Yang, 2009). However, little research on prospective teacher learning in China has been done. Based on the characteristics of secondary mathematics teacher preparation programs, Chinese prospective teachers are exposed to broad advanced mathematics content knowledge, some math education theories, and an extensive study on school mathematics, with little time devoted to student teaching (only around 4-6 weeks) (Li et al., 2008). Chinese pre-service teachers mainly obtain mathematics knowledge for teaching from their learning experience at pre-university and university.

With regard to mathematics classroom (pre-university) teaching in China, based on an extensive literature review, Huang and Li (2009) summarized the following features: (1) setting and achieving comprehensive and feasible teaching objectives; (2) having a detailed and well designed lesson plan that not only covers sufficient content to teach, but also offers alternatives to develop the content coherently; (3) emphasizing the formation and development of knowledge and mathematics reasoning; (4) emphasizing knowledge connection and instruction coherence; (5) practicing new knowledge with systematic variation problems; (6) making a balance between the teacher's guidance and students' self explorations; and (7) summarizing key points in due course and assigning homework (p.99).

Considering content coverage and content presentation in high school (Li, Zhang, & Ma, 2009), university teacher preparation programs as described above, the Chinese prospective teachers are more likely to develop sound subject content knowledge and a well-structured knowledge base. Based on this assumption, it is possible for Chinese prospective teachers to develop their fluency and flexibility simultaneously.

First, the teachers' sound content knowledge reduces cognitive load and leaves space for developing strategies in solving problems (Richland, Zur, & Holyoak, 2007). Second, the solid and interconnected knowledge base provides the foundations for developing flexibility. This implies that prospective teachers have a rich recipe of strategies for solving individual problems, and different representations for presenting mathematics concepts and mathematics problems. Third, it is a traditional and common practice to develop multiple approaches to solving a problem and developing multiple problems derived from the same problem or the same problem solving strategy in Chinese mathematics classrooms (Cai & Nie, 2007; Huang, Mok, & Leung, 2006). Implementing this approach of teaching requires learners to compare different strategies and select the most appropriate one for a particular type of problems. This comparison is an effective way to develop learners' flexibility in solving problems (Star & Seifert, 2006; Star & Rittle-Johnson, 2009).

In summary, the teacher preparation practice in China and learning experience in pre-university appears to provide opportunities for prospective teachers to develop their sound knowledge for teaching, and flexibility in selecting appropriate strategies and using appropriate representations of function. However, more empirical studies need to be done to explore relevant factors and mechanisms.

5.1.4 The Relationship between Prospective Teachers' KAT and Their Course Taking

This study revealed that the courses taken do have an effect on KAT, although the patterns in China and the U.S. differ. Chinese teacher preparation programs seem to emphasize content knowledge, while the U.S. teacher preparation programs put more emphasis on pedagogical knowledge. Meanwhile, the Chinese participants seem to have a more interconnected KAT structure than U.S. counterparts.

With regard to the number of courses taken, there was a big difference between China and the U.S. The average number of courses taken in mathematics content and mathematics education by the U.S. participants was 7 while the Chinese participants was 14. Thus, results of this study may be influenced by the difference of courses taken.

It is not surprising that there was such a big difference in the number of courses taken between China and the U.S.; in China, mathematics teachers for middle and high schools are required to major in mathematics. There was no distinction in preparing mathematics teachers for middle and high schools. Whether graduates work in high school or middle school depends on the job market and reputation of the university where they graduated. However, in the U.S., there are substantial differences between middle and high school mathematics teachers in terms of programs attended. In this study, the U.S. participants were from an interdisciplinary program of middle grade mathematics and science. It was suggested that it would be beneficial for middle and high school teachers to specialize in the subject field that they will be teaching (CBMS, 2001; National Commission of Mathematics and Science Teaching [NCMST], 2000). In particular, CBMS (2001) recommended middle grade mathematics teacher preparation programs should include at least 21 semester-hours of mathematics including two types of courses (as the participants in this study did). One focuses on developing a deep understanding of the mathematics they will be teaching. The other aims at strengthening and broadening understanding of mathematical connections between one educational level and the next. This includes connections between elementary and middle grades as well as between middle and high schools.

As far as algebra is concerned, the CBMS (2001) recommended developing a deep understanding of variables and functions as follows: (1) relate tabular, symbolic, and graphical representations to functions; (2) relate proportional reasoning to linear functions; (3) recognize change patterns associated with linear, quadratic, and exponential functions and their inverses; and (4) draw and use "qualitative graphs" to explore the meaning of graphs of functions. Meanwhile, students need to demonstrate the following skills: (1) represent physical situations symbolically; (2) graph linear, quadratic, exponential functions and their inverses and understand physical situations calling for each; (3) solve linear and quadratic equations and inequalities; and (4) exhibit fluency in working with symbols.

Comparing these recommendations in the U.S. and the practice in China, in order to implement curriculum standards (NCTM, 2000, 2006), the middle grade mathematics teacher preparation programs in the U.S. not only need to add more mathematics and mathematics education courses, but also need to improve the quality of courses and the quality of teaching.

5.2 Discussion

In the sections that follow, relevant issues that need to be further explored are discussed. These issues include measuring knowledge for teaching mathematics cross-culturally, comparison of MKT in different content areas, developing basic knowledge, skills and flexibility simultaneously, and what can be learned from this study.

5.2.1 Can Knowledge for Teaching Mathematics be Measured across Cultures?

Different models have been developed to define and measure mathematics knowledge for teaching. Although great efforts have been made to develop reliable instruments for measuring MKT in the U.S., there is not a well-developed instrument available at the present time. Even with the most popular one, developed by the University of Michigan, the low reliability of KCS (knowledge for content and student) (Schilling, Blunk, & Hill, 2007) inhibits it from being used in studies relating teacher knowledge to student achievement (Hill et al., 2004). Regarding the KAT instrument, although Floden et al. (2009) reported a high inter reliability rate (Cronbach alpha (α) .80 for whole instrument), the present study showed a relatively low reliability ($\alpha = 0.62$ for the U.S. sample ($N =115$) and $\alpha = .73$ for the Chinese sample ($N = 376$)).

In addition, the Chinese measurement model of KAT indicates that three components of KAT are highly correlated, and there are many links between different observed variables (items). It seems to suggest the complexity of mathematics knowledge for teaching: multiple venues for success and multiple solutions drawn on various knowledge and skills (Floden et al., 2009) results in the difficulty in measuring teacher knowledge for teaching by several isolated items. Ball, Thames, and Phelps (2008) realized that "it is not always easy to discern where one of our categories divides from the next, and this affects the precision (or lack thereof) of our definitions" (p. 403), and they also recognized that formulation of mathematical knowledge for teaching is culturally specific or dependent on teaching styles (Delaney et al., 2008). Based on specifying the relationship between Shulman's (1986, 1987) theorizations of "pedagogical content knowledge" and "the knowledge base for teaching," and Ball et al.'s (2008) notions of "specialized content knowledge" (SCK) and "mathematical knowledge for teaching," Lawrence (2010) concluded that Ball's model of "mathematical knowledge for teaching" would perhaps be made more useful for analyses of the kinds of knowledge-practice breakdown. However, if teacher knowledge should emphasize teachers' professional judgment as Shulman's, then the bridge building, namely "the wisdom of practice", is crucial. Furthermore, a shift in focus from mathematics teachers' knowledge to their knowledgeable practices might facilitate Ball's efforts to bridge professional knowledge and teaching practice (Lawrence, 2010).

Moreover, when developing an instrument for cross-culturally comparative studies of mathematical knowledge for teaching, the equivalence of coverage of items becomes a challenging issue. As pointed out by Delaney et al. (2008), factors such as teaching strategies, teacher beliefs, classroom contexts, the presence and prevalence of specific mathematical topics and the content of the textbooks should be considered. It was found that teaching strategies and emphases across different grades should be important factors influencing teachers' knowledge for teaching. For example, in the current study, when U.S. participants were asked why they did not connect algebraic and graphic representations when solving quadratic inequality, they said they were taught different methods by different teachers in different grades and they did not realize that these different types of knowledge are interconnected. Thus, it is important to consider teaching strategies across different grades.

In addition, the content placement across pre-university and university levels should be considered. For example, some content taught at the university level in the U.S. are the content in high school in China, and vice versa. In order to achieve common classifications of three components of KAT (*school math, advanced math and teaching math*), this content equivalence should be considered.

5.2.2 Do Chinese Teachers Have Sound Knowledge for Teaching?

This study showed that Chinese middle and high school prospective teachers demonstrated solid knowledge of algebra for teaching, with certain flexibility in adopting perspectives of function and selecting multiple representations of functions. Overall, Chinese prospective secondary mathematics teachers have a deeper understanding of mathematics knowledge of algebra for teaching compared with the U.S. counterparts.

This study extends Ma's (1999) investigation on elementary mathematics teachers' knowledge in China and the United States in some ways. The findings suggest that Chinese secondary mathematics teachers have a profound understanding of mathematical knowledge of algebra for teaching. In Ma's study, the focus was on the connection of relevant concepts, while the current study focused on the connections of different perspectives and representations of function, and the flexibility in selecting representations. This study also enriched and extended An et al. (2004) observations of pedagogical content knowledge of middle school mathematics teachers in China. An et al. found that the Chinese mathematics teachers emphasized gaining the correct conceptual knowledge by reliance on traditional, more rigid development of procedures. This study provides an explanation for why Chinese practicing teachers having a well-structured knowledge base obtained from their previous teacher preparation programs, and could emphasize gaining correct conceptual knowledge by reliance on "rigid development of procedures."

When considering the findings by Even (1998) and Black (2007), the strengths of Chinese mathematics teachers' knowledge of algebra for teaching are even more prominent. For example, 20% of 76 U.S. high school mathematics teachers in Blacks' (2007) study got correct answers, while 55% of 376 Chinese participants in the current study got correct answers. The finding that the Chinese secondary prospective mathematics teachers have sound mathematical knowledge of algebra for teaching may result from the secondary mathematics teacher preparation programs that emphasize mathematics content courses and mathematics education courses in China (Li et al., 2008).

5.2.3 Basic Knowledge and Skills and Flexibility in Problem Solving

If the core value of algebra learning is to develop students flexibility in translations between different representations and transformations between different forms within a presentation (Star & Rittle-Johnson, 2009) and to make sense of algebra through building connections between different brands of knowledge and different representations (NCTM, 2000, 2009), then, it is key to equip prospective teachers with relevant knowledge for teaching to promote these values.

Although mathematics education in China has a long tradition to pursue basic knowledge and basic skills of mathematics, mathematical thinking and rigorous logical reasoning (Zhang, Li, & Tan, 2004; Shao, Fan, Huang, Ding, & Li, 2013), great efforts to develop students' exploratory, collaborative learning, creative thinking, and mathematical communication have been made since 2001 (Ministry of Education, 2001, 2003). A study on comparing 10 novice and 10 expert teachers' views of effective mathematics (Huang, Li, & He, 2010) found that all the participating teachers valued students' mastering of mathematical knowledge and skills, and their development in mathematical thinking methods and abilities. Compared with novice teachers, expert teachers emphasized a greater emphasis on the development of students' mathematical thinking and higher order thinking abilities. These findings suggest that the mathematics teachers in the context of curriculum reform in China have been making a balance between mastering knowledge and skills and developing mathematics thinking and creative thinking.

This study revealed that the Chinese prospective teachers demonstrated sound knowledge and skills and fluency in algebra computation. In particular, the analysis of the open-ended items showed the Chinese prospective teachers not only had high fluency in algebraic computation, but also have flexibility in selecting appropriate function perspectives and using multiple representations. Moreover, the Chinese participants not only performed well, but also used various methods. Thus, the Chinese participants demonstrated high procedural fluency and a deep understanding of the concepts. It seems that the Chinese teachers develop their procedural fluency, conceptual understanding, and flexibility in adopting appropriate perspectives and selecting appropriate representations simultaneously.

5.2.4 What Can We Learn from the Study?

In the U.S., it is a publicly recognized problem that teachers do not have adequate knowledge of what they will be teaching. In this study, compared with Chinese counterparts, the weakness of the U.S. prospective teacher's knowledge of algebra for teaching is evident. Thus, I consider what U.S. mathematics educators may learn from Chinese teacher preparation practices.

First, adding more mathematics content courses to the existing teacher preparation programs may be necessary. Since the number of courses taken impact MKT and the U.S. participants took fewer courses than the Chinese participants, it is necessary to add more compulsory courses in mathematics and mathematics education in the U.S. middle grade mathematics teacher preparation program. Second, teaching approaches to mathematics education courses may also need to be improved. In China, the dominant teaching method is lecture with frequent probing questions which may be conducive to developing prospective teachers' systematic and interconnected knowledge, but may constrain their opportunities to develop creativity and inquiry. On the other hand, in the U.S., there are a lot of individual/on-line learning, student presentations, and projects in math education courses. These kinds of activities may be beneficial to developing individual exploration, team cooperation, and communication skills; however, prospective teachers may not be able to acquire necessary knowledge and skills systematically.

It may be important to ensure our prospective teachers having a deep understanding of core concepts and build an interconnected knowledge base through multiple approaches including direct instruction, problem-based teaching and learning, case studies, and inquiry projects.

In China, prospective teachers in this study demonstrated their strengths in mathematical knowledge of algebra for teaching and their flexibility in using appropriate representations. However, they need to learn more about how to develop students' creativity, discovery learning, and collaborative learning which are advocated in the new curriculum standards in China. Moreover, they need to learn how to develop a concept from multiple perspectives and how to build the connections between arithmetic, algebra, and geometry.

5.3   Limitation

There are several limitations in this study. The researcher considered the representativeness of the Chinese sample such as university entrance scores, programs, and regions, but the number of teacher education institutions in China is too large to be sampled by an individual research effort. In China, there was only one approach for preparing secondary school mathematics teachers (including middle school teachers) housed in mathematics departments. In the U.S., there are three approaches for preparing middle school mathematics teachers. The first prepares teachers to teach all secondary mathematics, including lower secondary/middle grades. The second focuses specifically and exclusively on preparing teachers for lower secondary/middle school grades. The third approach prepares lower secondary/ middle school teachers as an extension of elementary teacher preparation. In this study, the researcher mainly selected the participants from the second type of program in a respected public university (only a very small proportion of the participants from the first approach program). Because of this disparity, the sample at most reflects a low level of secondary math teacher preparation program in the U.S. Caution should be taken when interpreting the differences between China and the U.S. In order to make an appropriate comparison, wider samples from the different approaches in the U.S. should be included. Since there are essential differences of programs between China and the U.S., the numbers of courses taken are different. The number of courses taken in mathematics content and mathematics education, on average, was 7 by U.S. participants and 14 by Chinese participants. Caution should be taken in interpreting the findings of this study in which comparisons are made between U.S. and Chinese participants due to the disparity of the sample.

Second, the small size of the U.S. sample prohibited building a measurement model. If the U.S. sample size can be increased sufficiently to build structure equation models, then more sophisticated and extensive comparisons may be conducted.

Moreover, although the reliability (α = 0.88) for the whole sample (N = 491) is high, and the reliability (α = 0.73) for the Chinese sample (N = 376) is acceptable, but the reliability (α = 0.61) for the U.S. sample (*N* = 115) is relatively low. The results based on this instrument should be interpreted cautiously.

5.4 Recommendations

With regard to prospective teachers' mathematics knowledge for teaching, more questions need to be further explored. For example, what is meant by mathematics knowledge for teaching? How can it be measured? Do Chinese prospective teachers have sound mathematics knowledge for teaching in all areas of school mathematics? What strategies are effective in developing prospective teachers' basic knowledge and skills, and flexibility through teacher preparation program?

5.4.1 Study on the Meaning of Mathematics Knowledge Needed for Teaching

Concerning with the first question, many studies found the weaknesses of the existing instruments, such as the KAT in this study. This calls for researching into the nature of mathematical knowledge for teaching, and to what extent it can be measured. There are essentially two perspectives on the nature of MKT. The believers of cognitive perspective content advocate that MKT is related to "knowledge needed for teaching a specific subject" (Krauss et al., 2008, p. 716). However, the advocators of the situated cognitive perspective argue that MKT is a form of "knowledge in action" (Seymour & Lehrer, 2006). Building bridges between knowledge and practice toward a knowledgeable practice or wisdom of practice (Lawrence, 2010) shed light on defining and measuring teachers' knowledge needed for teaching.

5.4.2 Teachers' Knowledge for Teaching Other Topics in China and the U.S.

The second research question concerns specialty of topics investigated. Mathematics education in China has a tradition of emphasizing basic knowledge and skills (Zhang et al., 2004). Due to the core position of algebra in school mathematics, the teaching of algebra both at secondary schools and universities are emphasized. A great deal of time was spent on learning and practicing algebraic computation skills in high schools and universities. Prospective teachers in China may have strength in KAT, but it does not mean they should have strong knowledge for teaching other content areas, particularly in some newly added content such as probability and statistics. In the future, it will be interesting and meaningful to compare MKT in other content areas.

5.4.3 Are There Some Teaching Strategies Effective for Developing Flexibility?

Studies suggest that some strategies in China are effective for students learning with procedural fluency and conceptual understanding, such as problem-based teaching and practicing with varying problems (Gu, Huang, & Marton, 2004). Can these strategies be applied to teaching prospective teacher preparation courses in the U.S.? What are the effective teaching strategies for equipping prospective teachers with sound basic knowledge and skills, and flexibility in problem solving? Future studies should be interesting to explore how U.S. mathematics educators can adopt some Chinese strategies to the teacher preparation programs in the U.S. to develop teachers' knowledge needed for teaching.

5.5 Coda

A desire to find the similarities and differences of KMT and provide implications for improving mathematics teacher preparation in China and the U.S. led to this comparative study. Finally, the findings illustrated many more differences than similarities, and concluded the Chinese participants' superiority in knowledge of algebra for teaching. Although the findings may not be generalized due to the limitation of the bias of the sampling, the detailed description and analysis should provide references for international mathematics educators, particularly the Chinese and U.S. mathematics educators, to reflect upon what they can learn from this study. With less decisive statements, more meaningful research questions are generated for further investigations based on this study.

References

Alibali, M., Knuth, E., Hattikudur, S., Mcneil, N., & Stephens, A. (2007). A longitudinal examination of middle school students' understanding of the equal sign and equivalent Equations. *Mathematical Thinking and Learning, 9*, 221- 247.

An, S., Kulm, G., & Wu, Z. (2004). The pedagogical content knowledge of middle school mathematics teachers in China and the U.S. *Journal of Mathematics Teacher Education, 7*, 145-172.

An, S., Kulm, G., Wu, Z., Ma, F., & Wang, L. (2006). The impact of cultural differenece on middle school mathematics teachers' beliefs in the U.S. and China. In F. K. S. Leung, K. D., Graf, & F. J. Lopz-Real (Eds.), *Mathematics education in different cultural traditions: A comparative study of East Asia and the West* (pp.449-464). New York: Springer

Artigue, M., Assude, T., Grugeon, B., & Lenfant, A. (2001). Teaching and learning algebra: Approaching complexity through complementary perspectives. In H. Chick, K. Stacey, & J. Vincent (Eds.), *The Future of the teaching and learning of algebra (Proceedings of the 12th ICMI Study Conference* (pp. 21-32). Melbourne, Australia: The University of Melbourne.

Babcock, J., Babcock, P., Buhler, J., Cady, J. Cogan, L., Houang, R., et al. (2010). *Breaking the cycle: An international comparison of U.S. mathematics teacher preparations.* Michigan State University: The center for research in math and science education, Michigan State University.

Ball, D. L. (1990). The mathematical understandings that prospective teachers bring to teacher education. *The Elementary School Journal, 90*, 449-466.

Ball, D. L. (1993). With an eye on the mathematical horizon: Dilemmas of teaching elementary school mathematics. *The Elementary School Journal, 93*, 373-397

Ball, D. L., & Bass, H. (2000). Interweaving content and pedagogy in teaching and learning to teach: Knowing and using mathematics. In J. Boaler (Ed.), *Multiple perspectives on the teaching and learning of mathematics* (pp. 83-104). London: Ablex Publishing.

Ball, D. L., Hill, H. C., & Bass, H. (2005, Fall). Knowing mathematics for teaching: Who knows mathematics well enough to teach third grade, and how can we decide? *American Educator, 29*(3), 14-17, 20-22, 43-46.

Ball, D. L., Thames, M. H., & Phelps, G. (2008). Content Knowledge for Teaching: What Makes It Special? *Journal of Teacher Education, 59*, 389-407

Baumert, J., Kunter, M., Blum, W., Brunner, M., Voss, T. Jordan, A. et al. (2010). Teachers' mathematical knowledge, cognitive activation in the classroom, and student progress. *American Educational Research Journal, 47* (1), 133-180.

Bednarz, N., Kieran C., & Lee, L. (Eds.) (1996). *Approaches to algebra: Perspectives for research on teaching.* Dordrecht: Kluwer.

Black, D. J. W. (2007). *The relationship of teachers' content knowledge and pedagogical content knowledge in algebra and changes in both types of knowledge as a result of professional development.* Unpublished dissertation Auburn University, Auburn, Alabama.

Blömeke, S., & Delaney, S. (2012). Assessment of teacher knowledge across countries: A review of the state of research. *ZDM- The International Journal on Mathematics Education, 44*, 223-247.

Blume, G. W., & Heckman, D. S. (2000). Algebra and functions. In E. A. Silver & P. A. Kenney (Eds.), *Results from the seventh mathematics assessment of the national assessment of educational progress* (pp. 269-306). Reston, VA: National Council of Teachers of Mathematics.

Booth, L. R. (1984). *Algebra: Children's strategies and errors.* Windsor, UK: NFER-Nelson.

Brenner, M. E., Herman, S., Ho, H. Z., & Zimmer, J. M. (1999). Cross-national comparison of representational competence. *Journal for Research in Mathematics Education, 30*, 541–557.

Briedenbach, D. E., Dubinsky, E., Hawks, J., & Nichols, D. (1992). Development of the process conception of function. *Educational Studies in Mathematics, 23*, 247–285.

Buchholtz, N., Leung, F. K. S., Ding, L., Kaiser, G., Park, K., & Schwarz, B. (2013). Future mathematics teachers' professional knowledge of elementary mathematics from an advanced standpoint. *ZDM-International Journal on Mathematics Education, 45*, 107-120.

Byrne, B. (2010). *Structural equation modeling with AMOS: Basic concepts, applications, and programming* (2nd ed.). New York: Routledge.

Cai, J. (1995). *A cognitive analysis of U.S. and Chinese students' mathematical performance on tasks involving computation, simple problem solving, and complex problem solving.* Reston, VA: National Council of Teachers of Mathematics.

Cai, J. (2000). Mathematical thinking involved in U.S. and Chinese students' solving of process-constrained and process-open problems. *Mathematical Thinking and Learning, 2,* 309–340.

Cai, J. (2004). Why do U.S. and Chinese students think differently in mathematical problem solving? Impact of early algebra learning and teachers' beliefs *The Journal of Mathematical Behavior, 23,* 135-167.

Cai, J. (2005). U.S. and Chinese teachers' constructing, knowing and evaluating representations to teach mathematics. *Mathematical Thinking and Learning, 7,* 135–169.

Cai, J. (2006). U.S. and Chinese teachers' cultural values of representations in mathematics education. In F. K. S. Leung, K. D., Graf, & F. J. Lopz-Real (Eds.), *Mathematics education in different cultural traditions: A comparative study of East Asia and the West* (pp.465-482). New York, NY: Springer.

Cai, J., & Nie, B. (2007). Problem solving in Chinese mathematics education: Research and practice. *ZDM - The International Journal on Mathematics Education, 39,* 459-473.

Cai, J., & Wang, T. (2006). U.S. and Chinese teachers' conceptions and constructions of representations: A case of teaching ratio concept. *International Journal of Mathematics and Science Education, 4,* 145-186.

Cai, J., & Wang, T. (2010). Conceptions of effective mathematics teaching within a cultural context: perspectives of teachers from China and the United States. *Journal of Mathematics Teacher Education, 13,* 265–287.

Cai, J., Perry, B., Wong, N. Y., &Wang, T. (2009). What is effective teaching? Study of experienced mathematics teachers from Australia, the Mainland China, Hong Kong-China, and the United States. In J. Cai, G. Kaiser, B. Perry, & N. Wong (Eds.), *Effective mathematics teaching from teachers' perspectives: National and international studies* (pp.1-36). Rotterdam, The Netherlands: Sense.

Chazan, D., & Yerushalmy, M. (2003). On appreciating the cognitive complexity of school algebra: Research on algebra learning and directions of curricular change. In J. Kilpatrick, W. G. Martin, & D. Shifter (eds.),*A research companion to principles and standards for school mathematics*(p.123-136). Reston, VA : National Council of Teachers of Mathematics.

Chen, X., & Li, Y. (2010). Instructional coherence in Chinese mathematics class-room – a case study of lessons on fraction division. *International Journal of Science and Mathematics Education*. doi: 10.1007/s10763-009-9182-y

Chinnappan, M., & Thomas, M. (2001). Prospective teachers' perspectives on function representation. In J. Bobis, B. Perry, & M. Mitchelmore (Eds.) *Numeracy and beyond* (Proceeding of the 24 th annual conference of the Mathematics Education Research Group of Australisa, pp. 155-162). Sydney: MERGA.

Chrostowski, S. J., & Malak, B. (2004). Translation and cultural adaptation of the TIMSS 2003 instruments. In M. O. Martin, I. V. S. Mullis, & S. J. Chrostowski (Eds.), *TIMSS 2003 technical report* (pp.92-107) Chestnut Hill, MA: TIMSS & PIRLS International Study Center.

Common Core State Standards Initiative (CCSSI). 2010. *Common Core State Standards for Mathematics*. Washington, DC: National Governors Association Center for Best Practices and the Council of Chief State School Officers.
http://www.corestandards.org/assets/CCSSI_Math%20Standards.pdf.

Conference Board of the Mathematical Sciences. (2001). *The mathematical education teachers* (Vol. 4). Providence, RI and Washington DC: American Mathematical Society and Mathematical Association of America. Retrieved February 26, 2009, from http://www.cbmsweb.org/MET_Document/index.htm

Creenes, C. E., & Rubenstein, R. (2008). *Algebra and algebraic thinking in school mathematics: Seventeenth yearbook*. Reston, VA: The National Council of Teachers of Mathematics.

Creswell, J. W., & Clark, V. L. P. (2007). *Designing and conducting mixed methods research*. Thousand Oaks: Sage.

de Jong, T., Ainsworth, S., Dobson, M., van der Hulst, A., Levonen,J., & Reimann, P. (1998). Acquiring knowledge in science and math: The use of multiple representations in technology based learning environments. In M. W. van Someren, P. Reimann, H. P. A. Boshuizen, T. de Jong, et al. (Eds.), *Learning with multiple representations* (pp. 9–40). Amsterdam: Pergamon.

Delaney, S., Ball, D. L., Hill, H., Schilling, S. G., & Zopf, D. (2008). Mathematical knowledge for teaching: adapting U.S. measures for use in Ireland. *Journal of Mathematics Teachers Education, 11*, 171–197.

Depaepe, F., Verschaffel, L., & Kelchtermans, G.(2013). Pedagogical content knowledge: A systematic review of the way in which the concepts have pervaded mathematics educational research. *Teaching and Teacher Education, 34,* 12-25.

Doerr, H. M. (2004). Teachers' Knowledge and the Teaching of Algebra. In K. Stacey, C. Helen, & K. Margaret (Eds.), *The furture of the teaching and learning of algebra, The 12th ICM Study* (pp.267-290). Boston: Kluwer.

Dossey, J., Halvorsen, K., & McCrone, S. (2008). *Mathematics education in the United States 2008: A capsule summary fact book.* Reston, VA: National council of Teachers of Mathematics.

Dubinsky, E., & Harel, G. (1992). The nature of the process conceptions of function. In G. Harel & E. Dubinsky (Eds.), *The concept of function: Aspects of epistemology and pedagogy* (MAA Notes, Vol, 25, pp.85-106). Washington, DC: Mathematical Association of America.

Dubinsky, E., & Moses, R. P. (2011). Philosophy, Math Research, Math Ed Research, K-16 Education, and the Civil Rights Movement: A synthesis. *Notices of the American Mathematical Society, 58*(3), 401–409.

Dubinsky, E., & Wilson, R. (2013). High school students' understanding of the function concept. *The Journal of Mathematical Behavior, 32,* 83-101.

Edwards, E. L. (Ed.). (1990). *Algebra for everyone.* Reston, VA: National Council of Teachers of Mathematics.

Even, R. (1990). Subject matter knowledge for teaching and the case of functions. *Educational Studies in Mathematics, 21,* 521-544.

Even, R. (1993). Subject-matter knowledge and pedagogical content knowledge: Prospective secondary teachers and the function concept. *Journal for Research in Mathematics, 24,* 94-116.

Even, R. (1998). Factors involved in linking representations of functions. *Journal of Mathematical Behavior, 17,* 105-121.

Even, R.(1992). The inverse function: prospective teachers' use of "undoing". *International Journal of Mathematics Education in Science and Technology, 23,* 557-562.

Even, R., & Tirosh, D. (1995). Subject-matter knowledge about students as sources of teacher presentations of the subject-matter, *Educational Studies in Mathematics, 29,* 1-20.

Even, R., & Tirosh, D. (2008).Teacher knowledge and understanding of students' mathematical learning and thinking. In L. England (ed.), *Handbook of international research in mathematics education* (p.202-222). New York and London: Routledge, Taylor& Francis.

Fan, L., & Zhu, Y. (2007). Representation of problem-solving procedures: A comparative look at China, Singapore, and U.S. mathematics textbooks. *Educational Studies in Mathematics, 66*, 61-75.

Ferrini-Mundy, J., McCrory, R., & Senk, S. (2006, April). *Knowledge of algebra teaching: Framework, item development, and pilot results*. Research symposium at the research presession of NCTM annual meeting. St. Louis, MO.

Floden, R. E., & McCrory, R. (2007, January). *Mathematical knowledge of algebra for teaching : Validating an assessment of teacher knowledge.* Paper presented at 11th AMTE Annual Conference, Irvine, CA.

Floden, R. R., McCrory, R., Reckase, M. D.,& Senk, S. (2009, April). *Knowledge of algebra for Teaching: Validity studies of a new measure.* Paper presented at annual conference American Education Research Association, San Diego, CA.

Gagatsis, A., & Shiakalli, M. (2004). Ability to translate from one representation of the concept of function to another and mathematical problem solving. *Educational Psychology, 24*, 645–657.

Gess-Newsome, J. (1999). Introduction and orientation to examining pedagogical content knowledge. In J. Gess-Newsome & N. G. Lederman (Eds.), *Examining pedagogical content knowledge* (pp. 3–20). Dordrecht, the Netherlands: Kluwer.

Grossman, P. L., Wilson, S., & Shulman, L. (1989). Teachers of substance: Subject matter knowledge for teaching. In M. Renolds (ed), *Knowledge base for beginning teachers* (pp.23-36). New York, NY: Pergamon Press

Gu, M. (2006). The reform and development in teacher education in China. Keynote speech at the *First International Forum on Teacher Education.* Shanghai, China. Retrieved February, 4, 2008 from http://www.icte.ecnu.edu.cn/EN/show.asp?id=547

Hart, K. (Ed.) (1981). *Children's understanding of mathematics 11-16.* London: Murray.

Hiebert J., & Morris, A. K. (2009). Building a knowledge base for teacher education: An experience in K–8 mathematics teacher preparation. *The Elementary School Journal, 109*, 475-490.

Hiebert, J., Stigler, J., Jacobs, J., Givvin, K., Garnier, H., Smith, M., et al. (2005). Mathematics teaching in the United States today (and tomorrow): Results from the TIMSS 1999 Video Study. *Educational Evaluation and Policy Analysis, 27,*111–132.

Hill, H. C., Ball, D. L., & Schilling, S. G. (2004). Developing measures of teachers' mathematics knowledge for teaching. *The Elementary School Journal, 105*(1), 11-30.

Hill, H. C., Ball, D. L., & Schilling, S. G. (2008). Unpacking pedagogical content knowledge: Conceptualizing and measuring teachers' topic-specific knowledge of student. *Journal for Research in Mathematics Education, 39,* 372-400.

Hill, H. C., Blunk, M. L., Charalambous, C. Y., Lewis, J. M., Phelps, G. C., Sleep, L., & Ball, D. L. (2008). Mathematical knowledge for teaching and the mathematical quality of instruction: An exploratory study. *Cognition and Instruction, 26,* 430-511.

Hill, H. C., Rowan, B., & Ball, D. L. (2005). Effects of teachers' mathematical knowledge for teaching on student achievement. *American Education Research Journal, 42,* 371-406.

Hill, H. C., Schilling, S. G., & Ball, D. L. (2004). Developing measures of teachers" mathematics knowledge for teaching. *The Elementary School Journal, 105,* 12-30.

Hitt, F. (1994). Teachers' difficulties with the construction of continuous and discontinuous functions. *Focus on Learning Problems in Mathematics, 16*(4), 10-20.

Hu, L., & Bentler, P. M. (1999). Cutoff criteria for fit indexes in covariance structure analysis: Conventional criteria versus new alternatives. *Structural Equation Modeling, 6* (1), 1-55.

Huang, R, & Bao, J. (2006). Towards a model for teacher's professional development in china: Introducing keli. *Journal of Mathematics Teacher Education, 9,* 279-298.

Huang, R., & Cai, J. (2011). Pedagogical representations to teach linear relations in Chinese and U.S. classrooms: Parallel or hierarchical. *The Journal of Mathematical Behavior, 30, 149–165.*

Huang, R., & Cai, J. (2010). Implementing mathematics tasks in the U.S. and Chinese classroom. In Y. Shimizu, B. Kaur, R. Huang, & D., Clarke (Eds.),

Mathematical tasks in classrooms around the world (pp.147-166). Rotterdam: Sense

Huang, R., & Leung, F. K. S (2004). Cracking the paradox of the Chinese learners: Looking into the mathematics classrooms in Hong Kong and Shanghai. In L. Fan, N. Y. Wong, J. Cai, & S. Li (Eds.), *How Chinese learn mathematics: Perspectives from insiders* (pp.348-381). Singapore: World Scientific.

Huang, R., & Li, Y. (2009). Pursuing excellence in mathematics classroom instruction through exemplary lesson development in China: A case study. *ZDM - The International Journal on Mathematics Education, 41*, 297–309.

Huang, R., & Li, Y. (2010). Promoting mathematical understanding: An exploratory study of teaching algebra in U.S. and Chinese classrooms. In C., Keitel, K. Hino, R. Vithal, A. Begehr, & D. Clarke (Eds.), *Differences in mathematics classrooms internationally*. Rotterdam: Sense. (in press)

Huang, R., Li, Y., & He, X. (2010). What constitutes effective mathematics instruction: A comparison of Chinese expert and novice teachers' views. *Canadian Journal of Science, Mathematics and Technology Education.* (in press).

Huang, R., Li, Y., & Ma, T. (2010, October). Developing and mastering knowledge through teaching with variation: A case study of teaching fraction division. Paper to be presented at annual conference of the *North American Chapter of the International Group for the Psychology of Mathematics Education* (PME-NA), Columbus, Ohio.

Huang, R., Mok, I., & Leung, F. K. S. (2006). Repetition or variation: "practice" in the mathematics classrooms in China. In D. J. Clarke, C. Keitel, &Y. Shimizu (Eds.), *Mathematics classrooms in twelve countries: The insider's perspective* (pp.263-274). Rotterdam: Sense.

Huang, R., Ozel, Z. E. Y., Li, Y., & Rowntree, R. V. (2013). Does classroom instruction stick to textbooks? A case study of fraction division. In Y. Li. & G. Lappan (Eds.), Mathematics curriculum in school education (pp. 443-472). New York: Springer.

Kaput, J. J., Blanton, M. L., & Moreno, L. (2008). Algebra from a symbolization point of view. In J. J. Kaput, D. W. Carraher, & M. L. Blanton (Eds.), *Algebra in the early grades* (pp. 19–51). New York: Lawrence Erlbaum Associates.

Katz, V. J. (Ed.). (2007). *Algebra: Gateway to a technological future.* Washington, DC: The Mathematical Association of America.

Kieran, C. (1992). The learning and teaching of school algebra. In D. Grouws (Ed.), *Handbook of research on mathematics teaching and learning* (pp. 390-419). New York, NY: Macmillan Publishing Company.

Kieran, C. (2004). The core of algebra: Reflections on its main activities. In K. Stacey, C. Helen, & K. Margaret (Eds.), *The furture of the teaching and learning of algebra, The 12th ICM Study* (pp.35-44). Boston: Kluwer.

Kieran, C. (2007). Learning and teaching algebra at the middle school from college levels: Building meaning for symbols and their manipulation. In F. K. Lester, Jr. (Ed.), *Second handbook of research on mathematics teaching and learning* (pp.707-762). Charlotte, NC: Information age.

Kilpatrick, J., Blume, G., & Allen, B. (2006, May). *Theoretical framework for secondary mathematical knowledge for teaching.* Unpublished manuscript, University of Georgia and Pennsylvania State University. Available http://66-188-7644.dhcp.athn.ga.charter.com/Situations/%20ProposalDo cs/ProposDocs.html

Kilpatrick, J., Swafford, J., & Findell, B. (Eds.) (2001). *Adding it up: Helping children learn mathematics.* Washington, DC: National Academy Press.

Kline, R. B. (2005). *Principles and practice of structural equation modeling* (2nd). New York: Guilford Publications.

Krauss, K., Brunner, M., Kunter,M., Baumert, J. , Blum, W., Neubrand, M. et al. (2008). Pedagogical content knowledge and content knowledge mathematics teachers. *Journal of Educational Psychology,* 100, 716–725.

Krauss, S., Brunner, M., Kunter, M., Baumert, J., Blum, W., Neubrand, M., et al. (2008b). Pedagogical content knowledge and content knowledge of secondary mathematics teachers. *Journal of Educational Psychology, 100* (3), 716–725.

Kulm, G. (2008). A theoretical framework for mathematics knowledge in teaching middle grades. In G. Kulm (Ed.), *Teacher knowledge and practice in middle grades mathematics* (pp. 3-18). Rotterdam, The Netherlands: Sense.

Kulm, G., & Li, Y. (2009*)*. Curriculum research to improve teaching and learning: national and cross-national studies. ZDM-*The International Journal on Mathematics Education* , 41, 717-731.

Lawrence, A. M. (2010, May 3). From divides to bridges: A rhetorical perspective on mathematical knowledge for teaching. Presented at the annual conference of *American Educational Research Association*, Denver, CO.

Leinhardt, G., Zaslavsky, O., & Stein, M. K. (1990). Functions, graphs, and graphing: Tasks, learning, and teaching. *Review of Educational Research, 60,* 1–64.

Lesh, R., Post, T.,& Behr, M. (1987). Representations and translations among representing in mathematics learning and problems solving. In C. Janvier (1987), *Problems of representation in the teaching and learning of mathematics* (pp. 33-40). Hillsdale, NJ: Lawrence Erlbaum.

Leung, F. K. S. (1995). The mathematics classroom in Beijing, Hong Kong and London. *Educational Studies in Mathematics, 29,* 297-325.

Leung, F. K. S. (2005). Some characteristics of East Asian mathematics classroom based on data from the TIMSS 1999 Video Study. *Educational Studies in Mathematics, 60,* 199-215

Li, S., Huang, R., & Shin, Y. (2008). Mathematical discipline knowledge requirements for prospective secondary teachers from East Asian perspective. In P. Sullivan & T. Wood (Eds.), *Knowledge and beliefs in mathematics teaching and teaching development* (pp. 63-86). Rotterdam, The Netherlands: Sense.

Li, X. (2007). *An investigation of secondary school algebra teachers' mathematical knowledge of algebra for teaching ic equation solving.* Unpublished Doctoral Dissertation, University of Texas, Austin.

Li, Y., & Huang, R. (2008). Chinese elementary mathematics teachers' knowledge in mathematics and pedagogy for teaching: the case of fraction division. *ZDM - The International Journal on Mathematics Education, 40,*845–859.

Li, Y., & Huang, R. (Eds.) (2013). How Chinese teach mathematics and improve teaching. New York: Routledge Taylor & Francis Group.

Li, Y., Chen, X., & An, S. (2009). Conceptualizing and organizing content for teaching and learning in selected Chinese, Japanese and U.S. mathematics textbooks: The case of fraction division. *ZDM-The International Journal on Mathematics Education, 41,* 809-826.

Li, Y., Huang, R., & Yang, Y. (2010). Characterizing expert teaching in school mathematics in China: A prototype of expertise in teaching mathematics. In Y. Li & G. Kaiser (Eds.), *Expertise in mathematics instruction: An international perspective (pp. 167-195). New York: Springer.*

Li, Y., Zhang, J., & Ma, T. (2009). Approaches and practices in developing mathematics textbooks in China. *ZDM-The International Journal on Mathematics Education, 41,* 733-748.

Li, Y., Zhao, D., Huang, R., & Ma, Y. (2008). Mathematical preparation of elementary teachers in China: Changes and issues. *Journal of Mathematics Teacher Education, 11,* 417-430.

Llinares, S. (2000). Secondary School Mathematics Teacher's Professional Knowledge: a case from the teaching of the concept of function. *Teachers and Teaching: Theory and Practice, 6* (1), 41-62.

Lowery, N. V. (2002). Construction of teacher knowledge in context: Preparing elementary teachers to teach mathematics and science. *School Science and Mathematics, 102*(2), 68-83.

Ma, L. (1999). *Knowing and teaching elementary mathematics: Teachers' understanding of fundamental mathematics in China and the United States.* Mahwah, NJ: Erlbaum.

Magnusson, S., Krajcik, J. S., & Borko, H. (1999). Nature, sources and development of pedagogical content knowledge for science teaching. In N. G.Lederman (Ed.), *Examining pedagogical content knowledge: The construct and its implications for science education* (pp. 95-132). Netherlands: Kluwer.

Matz, M. (1982). Towards a process model for high school algebra errors. In D. Sleeman &J. S. Brown (Eds.), *Intelligent tutoring systems*(pp. pp. 25-50). New York: Academic Press

McCrory, R., Floden, R., Ferrini-Mundy, J., Reckase, M. D., Senk, S. L. (2012). Knowledge of algebra for teaching: A framework of knowledge and practices. *Journal for Research in Mathematics Education, 43* (5), 548-615.

Ministry of Education, P. R. China (2001a). *Agenda on the reform and development of the basic education by state department of P. R. China* [in Chinese]. Retrieved February, 6, 2008 from http://www.edu.cn/20010907/3000665.shtml.

Ministry of Education, P. R. China (1999). *Decision on deepening education reform and whole advancing quality education* [in Chinese].Retrieved February, 6, 2008 from http://www.edu.cn/20011114/3009834.shtml

Ministry of Education, P. R. China (1998). *An action plan for vitalizing education to face the 21 century* [in Chinese]. Retrieved February, 6, 2008 from http://www.moe.edu.cn/edoas/website18/level3.jsp?tablename=208&infoid =3337

Ministry of Education, P. R. China (2001b). *Mathematics curriculum standard for compulsory education stage (experimental version)* [in Chinese]. Beijing: Beijing Normal University Press.

Ministry of Education, P. R. China (2009). *Educational Statistics Yearbook of China 2008* Beijing: People's Education Press.

Monk, D. H. (1994). Subject area preparation of secondary mathematics and science teachers and student achievement. *Economics of Education Review, 13*, 125-145.

Moschkovich, J. , Schoenfeld, A. H., & Arcavi, A. (1993). Aspects of understanding: On multiple perspectives and representations of linear relations and connections among them. In Romberg, T. A., Fennema, E.,& Carpenter, T. P. (Eds.), *Integrating research on the graphical representation of functions* (pp. 69-100). Hillsdale, NJ: Lawrence Erlbaum.

Moses, R. P. (1995). Algebra, the new civil right. In C. B. Lacampagne, W. Blair, & J. Kaput (Ed.), *The algebra initiative colloquium* (vol. 2, pp. 53-67). Washington, DC: U.S. Department of Education, Office of Educational Research and Development.

Moses, R. P., & Cobb, C. E., Jr. (2001). *Radical equations: math literacy and civil rights*. Boston: Beacon.

National Assessment of Education Progress (2009). *National report card*. Retrieved April 10, 2010, from http://nationsreportcard.gov/

National Commission of Mathematics and Science Teaching for the 21st Century.(2000). *Before its' too late*. Washington, DC: Author.

National Council of Teachers of Mathematics. (2006). *Curriculum focal points for prekindergarten through grade 8 mathematic*. Reston, VA: NCTM

National Council of Teachers of Mathematics. (2000).*Principles and standards for school mathematics*, Reston, VA: Author.

National Council of Teachers of Mathematics. (2009). *Focus in high school mathematics: Reasoning and sense making*. Reston, VA: Author.

National Mathematics Advisory Panel. (2008). *Foundations for success: The final report of the National Mathematics Advisory Panel*. U.S. Department of Education: Washington, D.C.

National Middle School Association (NMSA).(2007). *Certification/Licensure by state*. Westerville, Ohio: NMSA, 2007. Retrieved April 10, 2010 from http://www.nmsa.org

Niess, M. L. (2005). Preparing teachers to teach science and mathematics with technology: Developing a technology pedagogical content knowledge. *Teaching and Teacher Education, 21*, 509-523.

Norman, A. (1992). Teachers' mathematical knowledge of the concept of function. In G. Harel & E. Dubinsky (Eds.); *The concept of function: Aspects of epistemology and pedagogy* (vol. 25, pp. 215-232). Washington, DC: Mathematical Association of America.

Norman, F. A. (1993). Integrating research on teachers' knowledge of function and their graphs. In Romberg, T. A., Fennema, E.,& carpenter, T. P. (Eds.), *Integrating research on the graphical representation of functions* (pp. 159-188). Hillsdale, NJ: Lawrence Erlbaum.

Organisation for Economic Co-operation and Development. (2006). PISA 2006 Technical Report. Paris, France: Author.

Piaget, J., & Moreau, A. (2001). The inversion of arithmetic operations (R. L. Campbell, Trans.). In J. Piaget (Ed.), *Studies in reflecting abstraction* (pp. 69–86). Hove, UK: Psychology Press.

RAND Mathematics Study Panel. (2003). *Mathematical Proficiency for All Students*. Santa Monica, CA: RAND

Richland, L. E., Zur, O., & Holyoak, K. J. (2007, may 25). Cognitive supports for analogies in the mathematics classroom. *Science, 316*, 1128-1129.

Robinson, K., Ninowski, J., & Gray, M. (2006). Children's understanding of the arithmetic concepts of inversion and associativity. *Journal of Experimental Child Psychology, 94*, 349–362.

Roth, W. M., & Bowen, G. M. (2001). Professionals read graphs: A semiotic analysis. *Journal for Research in Mathematics Education, 32*, 159–194.

Rowland, T., & Ruthven, K. (Eds.). (2010). Mathematical knowledge in teaching. Mathematics education library (Vol. 50). Berlin: Springer.

Schilling,S., Blunk, M., & Hill, H.C. (2007). Test validation and the MKT measures: Generalizations and conclusions. *Measurement, 5*(2–3), 118–128.

Schmidt, W. H., Tatto, M. T., Bankov, K., Blomeke, S., Cedillo, T., Cogan, L. et al. (2007). *The preparation gap: Teacher education for middle school mathe-*

matics in six countries. East Lansing, MI: Center for Research in Mathematics and Science Education, Michigan State University

Schwarts, J., & Yerushalmy, M. (1992). Getting students to function in and with algebra. In G. Harel & E. Dubinsky (Eds.), *The concept of function: Aspects of epistemology and pedagogy* (MAA Notes, Vol, 25, pp.261-289). Washington, DC: Mathematical Association of America.

Seymour, J. R., & Lehrer, R. (2006). Tracing the evolution of pedagogical content knowledge as the development of interanimated discourses. *Journal of the Learning Sciences, 15*, 549 - 582.

Shao, G., Fan, Y., Huang, R., Ding, E. & Li, Y., (2012). Examining Chinese mathematics classroom instruction from a historical perspective. In Y. Li & R. Huang (Eds.), *How Chinese teach mathematics and improve teaching* (pp.11-28). New York: Routledge.

Sfard, A. (1991). On the dual nature of mathematical conceptions: Reflections on processes and objects as different sides of the same coin. *Educational Studies in Mathematics, 22*, 1–36.

Sfard, A. (1992). Operational origins of mathematical objects and the quandary of reification: The case of function. In G. Harel & E. Dubinsky (Eds.), *The concept of function: Aspects of epistemology and pedagogy* (pp.59–84). Washington, DC: Mathematical Association of America

Sfard, A., & Linchevski, L. (1994) . The gains and the pitfalls of reification: The case of algebra. *Educational Studies in Mathematics, 26*, 191-228.

She, X., Lan, W., & Wilhlem, J. (2011). A comparative study on pedagogical content knowledge of mathematics teachers in China and the United States. *New Waves-Educational Research & Development, 14* (1), 35-49.

Shulman, L. S. (1986). Those Who Understand: Knowledge Growth in Teaching. *Educational Researcher, 15*(2), 4-14.

Silver, E. A., Meas, V. M., Morris, K. A., Star, J. R., & Benken, B. M. (2009). Teaching mathematics for understanding: An analysis of lessons submitted by teachers seeking NBPTS certification. *American Educational Research Journal, 46*, 501-531.

Silverman, J., & Thompson, P. W. (2008). Toward a framework for the development of mathematical knowledge for teaching. *Journal of Mathematics Teacher Education, 11*, 499–511.

Simon, M. (2006). Key developmental understandings in mathematics: A direction for investigating and establishing learning goals. *Mathematical Thinking and Learning, 8*, 359–371.

Smith, M. S., Arbaugh, F., & Fi, C. (2007). Teachers, the school environment, and students: Influences on students' opportunities to learn mathematics in grades 4 and 8. In P. Kloosterman & F. K. Lester Jr. (Eds.), *Results from the 2003 Assessment of the National Assessment of Educational Progress* (pp.191-226). Reston, VA: National Council of Teachers of Mathematics.

Star, J. R., & Seifert, C. (2006). The development of flexibility in equation solving. *Contemporary Educational Psychology, 31*, 280-300

Star, J., & Rittle-Johnson, B. (2009). Making algebra work: Instructional strategies that deepen students understanding, within and between representations. *ERS Spectrum, Spring 2009, 27 (2)*, 11-18.

Stein, M. K., Baxter, J. A., & Leinhardt, G. (1990). Subject-matter knowledge and elementary instruction: A case from functions and graphing. *American Educational Research Journal, 27*, 639-663.

Stein, M. K., Kaufman, J. H., Sherman, M., & Hillen, A. F. (2011). Algebra: A challenge at the crossroads of policy and practice. *Review of Educational Research, 81*, 453–492.

Stevenson, H. W., & Lee, S. (1995). The East Asian version of whole class teaching. *Educational Policy, 9*, 152–168.

Stevenson, H. W., Chen, C., & Lee, S. (1993). Motivation and achievement of gifted children in East Asia and the United States. *Journal for the Education of the Gifted, 16*, 223–250.

Stigler, J. W., & Hiebert, J. (1999). *The teaching gap: Best ideas from the world's teachers for improving education in the classroom.* New York: Free Press.

Stylianides, A. J., & Stylianides, G. J. (2006). Content knowledge for mathematics teaching: The case of reasoning and proving. In J. Novotna´, H. Moraova´, M. Kra´tka´, & N. Stehlikova´ (Eds.), *Proceedings of the 30th Conference of the International Group for the Psychology of Mathematics Education* (Vol. 5, pp. 201–208). Czech Republic: Charles University in Prague.

Tatto, M. T., Schwille, J., Senk, S. L., Ingvarson, L., Rowley, G., Peck, R., et al. (2012). *Policy, practice, and readiness to teach primary and secondary mathematics in 17 countries: Findings from the IEA Teacher Education and Development Study in Mathematics (TEDS-M).* Amsterdam: IEA.

Tatto, M. T., Schwille, J., Senk, S., Ingvarson, L., Peck, R., & Rowley, G. (2008). *Teacher education and development study in mathematics (TEDS-M): Policy, practice, and readiness to teach primary and secondary mathematics. Conceptual framework.* East Lansing, MI: Teacher Education and Development International Study Center, College of Education, Michigan State University.

Uesaka, Y., & Manalo, E. (2006). Active comparison as a means of promoting the development of abstract conditional knowledge and appropriate choice of diagrams in math word problem solving. In D. Barker-Plummer, R. Cox, & N. Swoboda (Eds.), *Proceedings of Diagrammatic representation and inference: 4th international conference, diagrams*(pp. 181–195). New York, NY: Springer.

Usiskin, Z. (1988). Conceptions of school algebra and uses of variables. In A. F. Coxford & A. P. Shulte(eds.), *Algebraic thinking, grades K–12* (pp. 8–19). Reston, VA: National Council of Teachers of Mathematics.

Verstappen, P. (1982). Some reflections on the introduction of relations and functions. In G van Barneveld & K. Krabbendam (Eds.), *Proceedings of Conference on Functions* (pp. 166-184). Enschede, The Netherlands: National Institute for Curriculum Development.

Vinner, S. (1983). Concept definition, concept image and the notion of function. *International Journal of Mathematics Education in Science and Technology, 14,* 239-305.

Wagner, S. M., Rachlin, S. L., & Jensen, R. J. (1984). *Algebra learning project: Final report.* Athens, GA: University of Georgia, Department of Mathematics Education..

Wagner, S., & Kieran, C. (1989). *Research issues in the learning and teaching of algebra.* Reston, VA: The National Council of Teachers of Mathematics.

Wang, T., & Murphy, J. (2004). An examination of coherence in a Chinese mathematics classroom. In L. Fan, N. Y. Wong, J. Cai, & S. Li (Eds.), *How Chinese learn mathematics: Perspectives from insiders* (pp.107-123). Singapore: World Scientific.

Whittington, D., (2002). Status of high school math teaching. Retrieved April 10, 2010 from http://2000survey.horizon-research.com/reports/high_math/high_math.pdf

Wood, T., Shin, S. Y., & Doan, P. (2006). Mathematics education reform in three US classrooms. In D. J. Clarke, C. Keitel, &Y. Shimizu (Eds.), *Mathematics classrooms in twelve countries: The insider's perspective* (pp.75-86). Rotterdam: Sense.

Yang, Y. (2009). How a Chinese teacher improved classroom teaching in Teaching Research Group: a case study on Pythagoras theorem teaching in Shanghai. *ZDM-International Journal on Mathematics' Educations, 41,* 279–296.

Yuan, Z. D. (2004). A transition from normal education to teacher education, *China Higher Education, 5,* 30–32 [in Chinese].

Zhang, D., Li, S., & Tang,R.(2004). The "Two Basics": mathematics teaching and learning in Mainland China. In L. Fan, N. Y. Wong, J. Cai, & S. Li (Eds.), *How Chinese learn mathematics: Perspectives from insiders* (pp.189-207). Singapore: World Scientific.

Zhou, Z., Peverly, S. T., & Xin, T. (2006). Knowing and teaching fractions: A cross-cultural study of American and Chinese mathematics teachers. *Contemporary Educational Psychology, 41,* 438-457.

Appendices

Appendix A: Rubrics

Item 18

Score	Description
4	Give the answers with the following elements: (a) Point out that (i) and (ii) are functions. (b) Point out that there is only one unique value corresponding to each value from domain value (such as one to one, multiple to one, but not one to multiple).
3	Give the answers with the following elements: (a) Point out that (i) and (ii) are functions; (b) The explanations do not relate to the key element (multiple to one or one to one), rather some superficial features such as: the function (i) with constant value, and the function (ii) is not continuous or expressed by two expressions or there are many holes.
2	(I): (i) (a) is correct: (i) and (ii) are functions. (b) without explanation or giving wrong explanation Or (II): (a) one of (i) and (ii) is a function, (b) give a correct explanation.
1	(I): (a) answer (i) is a function, (ii) is not, or is the inverse (b) explanation is missing or wrong. Or (II): (a) answer (i) and (ii) are not functions, but (b) give some relevant explanations.
0	Blank or total wrong answers in (i) and (ii)

Item 19

Score	Examples
4	1. Note that for ab to be positive either both a and b are positive, or both are negative. Solve and find x > 3 or x < -4.
	2. Solve x − 3 = 0 and x + 4 = 0; plot x = 3 and x = -4 on a number line. Identify whether (x-3)(x+4) is positive or negative on the three intervals determined by these 2 points.
	3. Rewrite (x-3)(x+4) and solve $x^2 + x - 12 = 0$ using quadratic formula. Then use one of Methods 1 or 2 above.
	4. Graph y = (x-3)(x+4). Identify x-values where parabola is above the x-axis.
3	1. Give two algebraic correctly.
	2. One method is correct while there are minor mistakes with the other.
2	1. There is only one correct method and solution.
	2. The two methods shown are essentially the same.
1	Based on the assumption that if ab is positive, then a and b are positive, namely, a > 0, b > 0. (Without using and, or or between two inequalities).
0	Blank or no mathematically useful statement

Item 20

Score	Explanation
4	Give correct answers with a counterexample.
3	Give correct answer with a correct counterexample, but minor calculation error or notational error or sloppy comment.
2	Give a correct judgment but do not provide relevant explanations.
1	There is at least one correct useful statement.
	For example, States YES or True but has something that might be relevant to the situation. For example, give examples such as A=0, then A△B=0 or B=0, then A△B=0.
0	Blank or useless information

Item 21

Score	Examples
4	Reason: stick to finding algebra expression and discriminant, without using graphical representation. Solution: According the given conditions, sketch a graph of $y = f(x) = ax^2 + bx + c$ and finding one root in $[1, 6]$. Moreover, according to the symmetry of a quadratic function, the quadratic function should have two intersection points at x-axis, namely, there are two roots of the $ax^2 + bx + c = 0$ ($a \neq 0$)
3	For example, just point out one root.
2	Just say students should consider by integrating numerical and pictorial representations.
1	Providing at least one related and useful statement
0	Blank or some information not related to solving this problem

Item 22

Score	Examples
4	Give answer C and provide different explanations such as: • Change of a leads change of the openness, thus a is not changed; the y-intercept is not changed, so c is not changed. Thus, it is only possible to change b. • The translated graph is the symmetrical graph of original graph with regard to y-axis. So b is changed into − b.
3	Answer C. However, the reason is not explained appropriately such as only mentioning a or c the invariance.
2	Give C or D and gives some explanations, with some serious mistakes, such as if a is changed then the graph is moved up or down.
1	Or give partly the features of graph when changing a,b, and c.

Item 23

Score	Examples
4	Use different forms to find the quadratic function. Find the maximum by using formula, symmetrical feature.
3	Find correct quadratic function expression but make mistakes in finding maximum.
2	Only find a correct quadratic function, without further attempt to find maximum.
1	Find one of a, b, c.

Item 24

Score	Example
	Method 1: Let $f(x)$ and $g(x)$ intersect at x-axis $(p, 0)$, then, $f(p) = 0$, $g(p) = 0$. So, $(f+g)(p) = f(p) + g(p) = 0 + 0 = 0$. Thus, $f+g$ $(p, 0)$
	Method 2: Let $f(x)$ and $g(x)$ intersect at x-axis $(p, 0)$, then, the following statements are true: (1) $f(p) = 0 \rightarrow ap + b = 0 \rightarrow p = -b/a$; (2) $g(p) = 0 \rightarrow cp + d = 0 \rightarrow p = -d/c$; (3) $f(p) = g(p) \rightarrow b/a = d/c \rightarrow ad = bc$; (4) $f(p) = g(p) \rightarrow ap + b = cp + d \rightarrow p = -(b + d)/(a + c)$; According to $(f+g)(p) = f(p) + g(p)$, and above statements, to deduce $(f+g)$ $(p) = 0$. Thus, $(f+g)(x)$ pass at point $(p, 0)$
3	All major points are made but one small piece may be skipped: Based on the above propositions (1) - (4), and deduce $(f+g)(x) = f(x) + g(x) = (a + c)x + (b + d)$, and get $(f+g)(p) = 0$, but make some minor mistakes.
2	Understand $f(p) = 0$, $g(p) = 0$, and $(f+g)(p) = f(p) + g(p)$, but they did not get $(f+g)(p) = 0$ or $(f+g)(x)$ passes at point $(p, 0)$. Although getting $(f+g)(x) = f(x) + g(x) = (a + c)x + (b + d)$, but fail to deduce $(f+g)(p) = 0$ by using previous propositions.
1	Understand f and g pass $P(p, 0)$, then, $f(p)= g(p) =0$, without further reasoning. Deduce $(f+g)(x) = f(x) + g(x) = (a + c)x + (b + d)$ without further reasoning.

Item 25

Score	Examples
4	Point out, if the x-axis represents time, and the y-axis represents the height above sea level, then origin explanation is correct.
	Other examples include: velocity vs. time, distance vs. time, or temperature vs. time, etc.
3	Point out it is not appropriate to describe the real situation without using mathematical relation (the meaning of x, and y).
	Gives a correct example, but does not provide details.
2	Student provides appropriate improving suggestion. Or gives a detailed example
1	Student points the inappropriate, such as direct description based on daily situation, or gives a piece of information about an example.

Appendix B: Interview Transcripts or Key Points

18. a) On a test a student marked both of the following as non-functions

    (i) $f: R \rightarrow R$, $f(x) = 4$, where R is the set of all the real numbers.

    (ii) $g(x) = x$ if x is a rational number, and $g(x) = 0$ if x is an irrational number.

    For each of (i) and (ii) above, decide whether the relation is a function, and write your answer in the Answer Booklet.

    b) If you think the student was wrong to mark (i) or (ii) as a non-function, decide   what he or she might have been thinking that could cause the mistake(s).

    Write your answer in the Answer Booklet.

    Questions: (1) How do you judge whether a relationship is a function or not?

    (2) What is vertical line test?

    (3) What would you teach to your students? Can you give an example?)

Larry: Larry gets a correct judgment, but does not provide any explanation and analysis of students' learning difficulties. When asking how to make his judgment, he says to use vertical line test and draw a diagram ($x = y^2$) and then judge it is not. He believes that students may be confused because of many holes, but the vertical line test can be passed.

Jenny: Jenny makes wrong answers without explanation. She has difficulty plotting the graph because she does not know how to calculate rational and irrational number appropriately. She says that she made her choice based on a visual image of the function. But she knows the vertical line test when giving a clue.

Kerri: Kerri clearly uses diagrams to explain the concept of function (one-to-one or multiple-to-one, but not one-to-multiple) and uses it to judge, as stated "a function is when one x value goes to one y, as long as one x value does not go to 2 y values, it is a function"

Alisa: Alisa clearly uses vertical line test. Although there are many holes in (ii), it still passes vertical line test. Students may be confused about horizontal line or vertical line test?

Stacy: She makes a correct judgment. Stacy states she used vertical line test. "Each input [value] should have only one [output] value, but that does not mean different inputs could not have different values". Students may be confused by the repeated output as stated "he could have seen outputs repeating and said non-function, even though there is actually one possible output for every input".

21. If you substitute 1 for x in expression $ax^2 + bx + c$ (a, b and c are real numbers), you get a positive number, while substituting 6 gives a negative number. How many real solutions does the equation $ax^2 + bx + c = 0$ have?

One student gives the following answer:

According to the given conditions, we can obtain the following inequalities:

$a + b + c > 0$, and $36a + 6b + c < 0$.

Since it is impossible to find fixed values of a, b and c based on the previous inequalities, the original question is not solvable.

Write down your answers in as much detail as possible on your Answer Booklet.

Questions: (1) What do you think about the reasons for the student's answers?

(2) To find solutions, what other mathematical concepts may be useful for solving the problem?

Larry: Larry totally agrees with students' comments. There are three parameters unknown. How can he find the solution to the equations? He tries to use algebraic transformation, but nothing can be done. He says he is stuck to algebraic operations; there is no idea to suggest how the student may go ahead. Even when the interviewer suggests drawing a graph, he still thinks it is impossible because all three coefficients are unknown. Even when the interviewer draws sketch of the quadratic function, he is not able to build the connections between roots and intersection points.

Jenny: Jenny honestly says that she just jumped to the conclusion. "I actually do not know how to solve this problem at all". What she could suggest is to ask students to try different ways, such as plugging in more numbers between 1 and 6. She directly asks the interviewer if he can tell a method. When the interviewer asks her to read the question carefully, to understand the question is to find the number of roots. The interviewer gives hints: if you have difficulty in thinking algebraically, can you consider in other representations? Can you use graphical representations? Then the interviewer draws a sketch of the quadratic function based on the given conditions. Then the interviewee was enlightened to think roots and intersect points. She is not quite sure, but she finally finds the number of roots. However, she says she did not have this experience in solving inequalities using graphing.

Kerri: Kerri agrees with the student's statement. She has no ideas how to find the results of the questions. They tried to find a, b, and c by adopting the ideas from algebra. But it does not work. Even when hinting to draw a graph based on the given conditions, she still gets stuck because she does not know a, b and c. When the interviewer draws the sketch, she realizes there are two roots.

Alisa: Honestly, she does not know how to solve the question. She agrees with students. I do not know how to solve this problem. When asking whether she can fix out the number of roots by other methods, she guesses she can. She draws a sketch of a quadratic function, and found the number of roots (two). She realizes the power of graphing method and will teach her students.

Stacy: She does not have clear ideas about how to solve this problem. But she would like to suggest students to explore different ways such as plugging in more numbers to see whether they can find patterns, rather than being stuck. However, she still intends to find out a, b and c. When enlightening whether graphing method can be used, she draws a correct graph, and find the number of roots.

22. Mr. Seng's algebra class is studying the graph of $y = ax^2 + bx + c$ and how changing the parameters a, b, and c will cause different translations of the original graph.

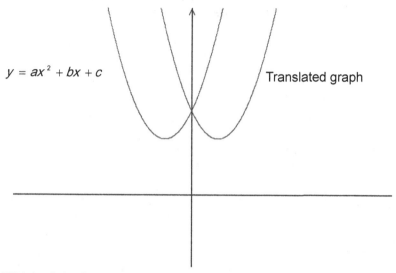

$y = ax^2 + bx + c$ Translated graph

Which of the following is an appropriate explanation of the translation of the original graph $y = ax^2 + bx + c$ to the translated graph?

    A. Only the a value changed.
    B. Only the c value changed.
    C. Only the b value changed.
    D. At least two of the parameters changed.
    E. You cannot generate the translated graph by changing any of the parameters.

Explain your answer choice in as much detail as possible. Show your work in the Answer Booklet.

Questions: (1) What are the effects of changing parameters of a, b, c
               on the change of graph?
               (2) What do you find the variance or invariance after the
               translations?
               (3) What algebraic operations do you think useful for
               identifying the key parameter(s)?

Larry: If c is changed, the graph is moved up or down; if b is changed,
the graph is moved from left to right or reverse; if a is changed,
the shapes of the graph is changed; so only b can be changed
(correct answer should be C)

Jenny: She did a lot of translations of graphs in high school: b -> -b just
influence on the left and right. But she is not clear about the effect of changing a, b, c. It is a good way to show students different examples to explain the effect of a, b, and c.

Kerri: She understands that changing c causes moving up or down of
the graph; changing b causes the graph to be wider or narrower;
thus, only b can be changed. (Just make wrong choice).

Alisa: She understands that changing c results in moving up or down of
the graph; changing b results in expanding or compressing the
graph. But she makes mistakes in drawing a graph. So, she
makes a wrong choice. She does not know how to translate
graph. (for example, from f(x) to f(x+h)).

Stacy: She knows how changes of a and c impact on the changes of
the graph. But she is not clear about the change of b and its impact. Solution is correct. But she is not sure why! She realizes
that she learned them in algebra I or II classes.

23. When introducing the functions and the graphs in a class of middle school (14-15 year-olds), tasks were used which consist of drawing graphs based on a set of pairs of numbers contextualized in a situation and from equations. One day, when starting the class, the following graph was drawn on the blackboard and the pupils were asked to find a situation to which it might possibly correspond.

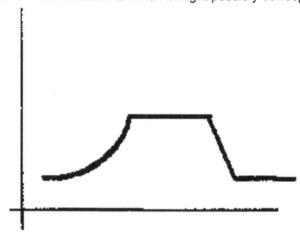

One student answered: 'it may be the path of an excursion during which we had to climb up a hillside, the walk along a flat stretch and then climb down a slope and finally go across another flat stretch before finishing.'

How could you respond to this student's comment? What do you think may be the cause of this comment? Can you give any other explanations of this graph?

Write down your answers in as much detail as possible on your Answer Booklet.

Questions: (1) What are the missing parts of the students' comment (two variables, X vs. Y);

(2) How can you explain other real life situations using this graph?

Larry: He draws a diagram with x-axis (time) and y-axis (position), and explains how this diagram makes sense of the situation. He also gives two examples of Speed vs. Time and Temperature vs. Time.

Jenny: Jenny says it could be speed over time. In one course of math and technology, she learned this kind of graph through experiment.

Kerri: She also talks about speed over time. She gave a detailed description as to increase, constant, decrease and constant.

Alisa: She also understands the graph as distance over time, or speed over time. The key is the relationship between two variables of x and y.

Stacy: She explains that the X-axis represents time, while the Y-axis represents distance, namely, height. She describes the changing trends such as flat, growing up, and going down. In general, it could represent the distance vs. time or *temperature vs. time*.

24.   Given quadratic function $y = ax^2 + bx + c$ intersects x-axis at (-1, 0) and (3, 0), and its y-intercept is 6.   Find the maximum of the quadratic function.

Show your work in as much detail as possible in the Answer Booklet.

Questions:

(1)   What kinds formula of the quadratic functions do you prefer to use? (how many different formulas have you learned) ?
(2)   To find maximum, which formula do you prefer?

Larry: He knows the methods: plugging three points into quadratic equations to find a, b and c, and then taking the derivative to find the turning point where the y-value is the maximum. He also points out that the graph should be symmetric with respect to line x=1, and when x=1 the y value should be the maximum. But he does not know any other forms of quadratic function.

Jenny: She has difficulty finding a, b, and c.   She guesses that y-intercept is the maximum. However, when drawing a sketch of quadratic equation, she realizes that the maximum should be reached at x=1.

Kerri: She supposes that the y-intercept is the maximum. She just draws three points and tries to see the maximum. She realizes the method, but cannot remember the formula clearly.

Alisa: She finds the expression of quadratic function (in general form) by plugging in three coordinates of points. And then using the symmetry of quadratic graph, she realizes that when x=1, y=8 is the maximum. However, she does not have any ideas about the other forms, which could be used to solve this problem.

Stacy: She makes mistakes in plugging in x values in order to find a, b, and c. But she knows that she can find the maximum by taking the derivate. She does not realize there are other forms of quadratic functions.

25. Prove the following statement:

If the graphs of linear functions $f(x) = ax + b$ and $g(x) = cx + d$ intersect at a point P on the x-axis, the graph of their sum function $(f + g)(x)$ must also go through P.

Show your work in as much detail as possible in the Answer Booklet.

Q: (1) What does mean by intersect at a point on x-axis?

(2) What is the mean of (f+g)(x)?

(3) What do you want to prove?

Larry: In his answer to questionnaire, he used two concrete examples to computation. However, during the interview, he used the general form and find correct proof as the following: $ax_1+b=0$, $ax_2+b=0$; $(f+g)(x)=(ax_1+b)+(ax_2+b)=0+0=0$;

Jenny: She just gives two concrete examples, and shows how to find the intersection points. Then she fails to show how to prove.

Kerri: She gave up the question in questionnaire.

Alisa: She gave up the question in questionnaire.

Stacy: She realizes how to use visual methods. She does not like proving, and has no idea how to prove it.

Appendix C: SME Model Parameters

Table C1.: Selected AMOS Output for the Final Chinese Model: Unstandardized and Standardized Estimate

			Estimate	S.E.	C.R.	P
			Regression Weights			
MKT3	<---	SM	.045	.019	2.358	.018
MKT6	<---	SM	.168	.050	3.359	***
MKT14	<---	SM	.114	.034	3.309	***
MKT10	<---	TM	.115	.034	3.358	***
MKT25	<---	TM	.863	.119	7.280	***
MKT18	<---	TM	.723	.096	7.567	***
MKT8	<---	AM	.234	.058	4.066	***
MKT9	<---	AM	.213	.079	2.677	.007
MKT12	<---	AM	.273	.084	3.269	.001
MKT13	<---	AM	.181	.075	2.406	.016
MKT16	<---	AM	.296	.060	4.924	***
MKT20	<---	AM	1.000			
MKT4	<---	AM	.308	.076	4.035	***
MKT17	<---	SM	.074	.020	3.768	***
MKT24	<---	AM	1.498	.305	4.907	***
MKT22	<---	TM	.739	.120	6.151	***
MKT21	<---	TM	1.000			
MKT23	<---	SM	1.000			
MKT19	<---	SM	.916	.124	7.410	***
MKT2	<---	PCK1	.048	.014	3.433	***
MKT7	<---	TM	.089	.028	3.190	.001

	Estimate	S.E.	C.R.	P

Standardized Regression weights

			Estimate
MKT3	<---	SM	.147
MKT6	<---	SM	.215
MKT14	<---	SM	.211
MKT10	<---	PCK1	.206
MKT25	<---	PCK1	.522
MKT18	<---	PCK1	.544
MKT8	<---	AM	.341
MKT9	<---	AM	.185
MKT12	<---	AM	.237
MKT13	<---	AM	.163
MKT16	<---	AM	.474
MKT20	<---	AM	.374
MKT4	<---	AM	.323
MKT17	<---	SM	.243
MKT24	<---	AM	.488
MKT22	<---	PCK1	.420
MKT21	<---	PCK1	.579
MKT23	<---	SM	.530
MKT19	<---	SM	.677
MKT2	<---	PCK1	.211
MKT7	<---	PCK1	.196

			Estimate	S.E.	C.R.	P
			Covariances			
SM	<-->	PCK1	.469	.078	6.020	***
PCK1	<-->	AM	.299	.062	4.851	***
SM	<-->	AM	.200	.045	4.432	***
			Correlations			
SM	<-->	PCK1	.908			
PCK1	<-->	AM	.827			
SM	<-->	AM	.747			

Table C2.: Selected AMOS Output for Final Chinese Model: Goodness-of-Fit Statistics

Model fit Summary

CMIN

Model	NPAR	CMIN	DF	P	CMIN/DF
Default model	57	245.347	174	.000	1.410
Saturated model	231	.000	0		
Independence model	21	1036.888	210	.000	4.938

RMR, GFI

Model	RMR	GFI	AGFI	PGFI
Default model	.027	.943	.925	.711
Saturated model	.000	1.000		
Independence model	.146	.696	.666	.633

Baseline comparison

Model	NFI Delta1	RFI rho1	IFI Delta2	TLI rho2	CFI
Default model	.763	.714	.917	.896	.914
Saturated model	1.000		1.000		1.000
Independence model	.000	.000	.000	.000	.000

Model fit Summary

CMIN

Model	NPAR	CMIN	DF	P	CMIN/ DF

RMSEA

Model	RMSEA	LO 90	HI 90	PCLOSE
Default model	.033	.023	.042	.999
Independence model	.102	.096	.109	.000

AIC

Model	AIC	BCC	BIC	CAIC
Default model	359.347	366.452	583.334	640.334
Saturated model	462.000	490.793	1369.735	1600.735
Independence model	1078.888	1081.506	1161.409	1182.409

ECVI

Model	ECVI	LO 90	HI 90	MECVI
Default model	.958	.858	1.079	.977
Saturated model	1.232	1.232	1.232	1.309
Independence model	2.877	2.619	3.155	2.884

Table C3.: Selected AMOS Output for the Final American path analysis Model: Unstandardized

			Estimate	S.E.	C.R.	P
			Regression Weights			
SM	<---	N_Highmath	.284	.212	1.342	.180
SM	<---	N_Collemath	.143	.063	2.277	.023
AD	<---	N_Collemath	.070	.069	1.014	.311
AD	<---	Grade	.030	.221	.137	.891
AD	<---	SM	.253	.101	2.519	.012
TM	<---	N_Highmath	.195	.408	.478	.633
TM	<---	Grade	-.178	.386	-.461	.645
TM	<---	N_Collemath	.247	.124	1.991	.047
TM	<---	AD	.422	.161	2.620	.009
TM	<---	SM	.769	.181	4.244	***
SM	<---	N_Highmath	.284	.212	1.342	.180
SM	<---	N_Collemath	.143	.063	2.277	.023
AD	<---	N_Collemath	.070	.069	1.014	.311
AD	<---	Grade	.030	.221	.137	.891
AD	<---	SM	.253	.101	2.519	.012
TM	<---	N_Highmath	.195	.408	.478	.633
TM	<---	Grade	-.178	.386	-.461	.645
TM	<---	N_Collemath	.247	.124	1.991	.047
TM	<---	AD	.422	.161	2.620	.009
TM	<---	SM	.769	.181	4.244	***
SM	<---	N_Highmath	.284	.212	1.342	.180

			Estimate	S.E.	C.R.	P
		Covariances				
N_Highmath	<-->	N_Collemath	.472	.153	3.080	.002
N_Collemath	<-->	Grade	-.272	.151	-1.798	.072
N_Highmath	<-->	Grade	-.037	.044	-.839	.401

Table C4:   Selected AMOS Output for Final American Path Analysis
Model: Goodness-of-Fit Statistics

Model Fit Summary

CMIN

Model	NPAR	CMIN	DF	P	CMIN/DF
Default model	19	.079	2	.961	.040
Saturated model	21	.000	0		
Independence model	6	73.016	15	.000	4.868

RMR, GFI

Model	RMR	GFI	AGFI	PGFI
Default model	.007	1.000	.998	.095
Saturated model	.000	1.000		
Independence model	.915	.796	.714	.568

Baseline comparison

Model	NFI Delta1	RFI rho1	IFI Delta2	TLI rho2	CFI
Default model	.999	.992	1.027	1.248	1.000
Saturated model	1.000		1.000		1.000
Independence model	.000	.000	.000	.000	.000

Model Fit Summary

CMIN

Model	NPAR	CMIN	DF	P	CMIN/DF

RMSEA

Model	RMSEA	LO 90	HI 90	PCLOSE
Default model	.000	.000	.000	.971
Independence model	.182	.141	.225	.000

AIC

Model	AIC	BCC	BIC	CAIC
Default model	38.079	40.497	90.722	109.722
Saturated model	42.000	44.673	100.184	121.184
Independence model	85.016	85.779	101.640	107.640

ECVI

Model	ECVI	LO 90	HI 90	MECVI
Default model	.325	.342	.342	.346
Saturated model	.359	.359	.359	.382
Independence model	.727	.530	.987	.733

Table C5: Selected AMOS Output for the Final Chinese path analysis
Model: Unstandardized

			Estimate	S.E.	C.R.	P
Regression Weights						
SM	<---	Grade	-.919	.204	-4.510	***
AD	<---	N_Collemath	.098	.044	2.236	.025
AD	<---	Grade	.423	.288	1.472	.141
AD	<---	SM	.583	.061	9.634	***
TM	<---	AD	.437	.075	5.823	***
TM	<---	SM	.807	.098	8.217	***
TM	<---	N_Collemath	.066	.064	1.027	.305
TM	<---	Grade	-.489	.419	-1.167	.243
Covariances						
N_Collemath	<-->	Grade	.846	.093	9.112	***